Intermediate Tropical Agriculture Series

General Editor

W.J.A. Payne,
Consultant in Tropical Livestock Production

Tropical Pastures and Fodder Crops

L.R. Humphreys M.Sc.Agr., Ph.D.

Department of Agriculture
University of Queensland, Australia.

Longman Group Limited
Longman House
Burnt Mill, Harlow, Essex, CM20 2JE, England.

First published 1978
Third impression 1982
ISBN 0 582 60303 X

Text set in 10/11 pt Photon Times,

Printed in Hong Kong by
Wing Tai Cheung Printing Co Ltd

Acknowledgements

The publishers are grateful to the following for permission to reproduce copyright material:

Commonwealth Scientific & Industrial Research Organisation for various tables from past issues of *Australian Journal of Experimental Agriculture & Animal Husbandry* and *Australian Journal of Agricultural Research*; Netherlands Journal of Agricultural Science for a table by Boonman & van Wijk in the *Netherlands Journal of Agricultural Science*, 21 (1973) 12; Tropical Grasslands Society of Australia for various tables from past issues of *Tropical Grasslands*.

We regret that we have been unable to trace the copyright holders of some of the data included in this book and would appreciate any information that would enable us to do so.

To I.L.H.

Other Titles in the Intermediate Tropical Agriculture Series

Already published:

H. T. B. Hall; *Diseases and Parasites of Livestock in the Tropics.*
Describes the causes, symptoms, treatment and control of the main diseases of livestock in the tropics.

To be published 1978:

J. C. Abbott; *Agricultural Economics and Marketing in the Tropics.*
A comprehensive and practical account of pricing, marketing, farm management and development economics.

Other titles in preparation include the following:

C. N. Williams & W. Y. Chew; *Tree and Field Crops of the Wetter Regions of the Tropics.*
D. S. Hill & J. M. Waller; *Principles and Methods of Pest and Disease Control of Tropical Crops.*
J. A. Eusebio; *Pig Production in the Tropics.*

Contents

PART 4: PASTURE ESTABLISHMENT

PART 5: SOIL FERTILITY AND FERTILISER NEEDS OF TROPICAL PASTURES

PART 6: PASTURE MANAGEMENT AND PRODUCTIVITY

Glossary

aggregation Uniting of soil particles.
Al Aluminium.
anaerobic Without oxygen.
anion Negatively charged ion.
anthesis Discharge of male pollen.
apical Of the top.
apomictic Producing seeds of identical genetic make-up to the mother plant.
aspect Direction which sloping ground faces.
assimilate Combined carbon formed by photosynthesis and moved in the plant.
auxins Growth regulating hormones of plants.
axillary From the upper angle between leaf and stem.
B Boron.
basaltic derived soils Soils developed from basalt rock.
biomass Weight of living organisms per unit area.
bituminous Made from a tar-like substance called bitumen.
botanical composition (percentage) Proportion of pasture made up by different botanical species.
bunch grasses Grasses forming a separate tussock with a clear boundary around the plant.
bush-fallow system A system of retiring land from cultivation.
Ca Calcium.
canopy Aerial parts of the pasture, especially the leaves.
catchment Area which collects rainfall.
cation Positively charged ion.
chernozem Black earth soils having a high clay content which changes little with depth; usually neutral or alkaline
chlorophyll Compound involved in photosynthesis.
chloroplast Granule containing chlorophyll.
chlorosis Blanching of green tissues.
climax Culmination of a series of changes in botanical composition.
Co Cobalt.
coenzyme Substance activating or accelerating enzyme action.
companion Associated.
complementary Serving to complete or to balance a deficiency.
control Standard for comparing effect of treatments.
cracking clay Clay which forms cracks upon drying.
crown Group of central basal stems.
crown roots Roots arising directly from the crown.
culms Stem.
cultivar Named botanical variety.
Cu Copper.
cv. Cultivar.
cytochromes Proteins essential for oxidation-reduction processes in cells.
dam catchment Catchment area above a water storage reservoir.
decumbent Lying along the ground.
defoliation Removal of shoots.
development phase Stage of development or progress to reproduction.
disclimax Stage in plant succession which replaces a true climax because of some disturbance, e.g. fire.
distal Away from point of attachment.
draught animals Animals used to provide power.
ecosystem System formed by the interaction of organisms and their environment.
ecotype A plant type from a particular environmental habitat.
edaphic Influenced by soil conditions.
enzyme Catalyst produced by living organisms.
establishment Sowing or planting a new pasture.
fallowing Keeping land uncropped.
Fe Iron.
genetic base Genetic constitution.
genetic shift Change in genetic constitution.
germinability Capacity to germinate or form seedling from seed.
glabrous Hairless.
gleyed Waterlogged, usually with grey or blue-grey sub-soil.
granitic Derived from granite.
grubbing Digging out roots.
heath Low vegetation with shrubs.
herbage Low growing plants used for fodder.
hessian Cloth made from hemp or jute.
humic Moist.
hydrolytic Causing hydrolysis or the reaction between a compound and the hydrogen and hydroxyl ions.
imbibition Absorption of water.
indeterminate Not fixed in number.

inflorescence Structure bearing flowers.
interveinal Between the structural vessels or veins.
in vitro Used of digestibility estimates made in laboratory with rumen liquor.
in vivo In the live animal.
isotherm Line joining points on the earth's surface having equal temperatures.
K Potassium.
krasnozems Are red, well-structured clay soils; clay content increases with depth.
labile Liable to change.
LAI Leaf area index, the area of green plant surface supported by unit ground area.
lamina Leaf blade.
latosol See krasnozem.
leach To dissolve substances with percolating water.
leaf area index See LAI.
ley Temporary pasture in rotation with crops.
lodging Crop bent over close to ground.
membrane Thin film.
meristem Growing point.
mesophyll Soft wood.
metabolism Chemical processes of degradation and synthesis in living organisms.
Mg Magnesium.
micro-elements Elements occurring in small amounts.
Mn Manganese.
Mo Molybdenum.
monogastric Single stomached.
monospecific pastures Pastures composed of a single species.
montane Mountainous.
N Nitrogen.
necrotic Dead.
night yarding Putting animals in a confined space at night.
node Joint of a stem at which leaves arise.,
oestrogenic Used of substances inducing oestrus or sexual heat.
optimum Most favourable.
overheads Costs.
overseed Broadcasting seeds on the soil surface.
P Phosphorus.
palatable Well eaten by animals.
perennial Living several years.
petiole Stalk supporting leaf blade.
pH Degree of acidity; >7 alkaline, <7 acid.
photoperiod Duration of daylight.
photosynthate Product of photosynthesis, where sun energy fixes carbon dioxide as plant carbohydrate.
plane of nutrition Level of feeding.
plant habit Plant structure or erectness.
podzolic Used of acid sandy soils with contrasting red, yellow or black subsoil.
promiscuous Used of legume root-nodule bacteria: forming nodules with many local strains.
quiescent Not growing.
radiation frost Frost caused by the ground radiating or losing heat at night.
respiration Plant process involving oxygen absorption and carbon dioxide emission.
rhizobial Of the legume root nodule bacteria which fix nitrogen from the atmosphere.
runnering With creeping stems or roots.
runoff Water flowing on the ground surface.
S Sulphur.
scarification Scratching or changing the seed coat.
scrambling legumes Legumes with a climbing or trailing habit.
sclerophyll Hardwood.
sedge A plant of the family Cyperaceae.
seed flow Movement of seed through a machine.
seminal Seedling.
senescent Growing old.
seral Successional stage in plant communities.
shading Reducing sunlight received by a plant.
short day (obligate) Plants which will not flower if days are longer than a certain amount.
siliceous Sandy.
sod Layer of basal stems.
solodic Soils with strong differences in texture with increasing depth; acid surface layers and strongly alkaline deeper layers.
solonised solonetz Similar to a solodic soil but with soil columns developed in the lower horizons.
solonchak A highly saline soil type.
spelling Resting a pasture from grazing.
sprig Portion of stem.
standing feed Pasture standing in the field.
stand over Surplus pasture standing in the field.
stolon Decumbent stem.
stooling out New shoots appearing on the edge of the crown.
stress Constraint upon plant growth or survival.
stylo Common name for legume *Stylosanthes*.
symbiosis Mutually beneficial association between two

life forms.

tiller Grass shoot.

topography Features of the natural land surface.

transpiration Evaporation of water from leaf surface.

tubercle Rounded protuberance.

understory Layer of plants near to the ground.

ungulate Hoofed animal.

varietal stability Degree of change and variation occurring in a cultivar or variety.

viability Capacity to germinate.

volatilisation Formation of gases.

voiding Evacuating.

volunteer Species which regenerates without being sown.

windrowing Pushing plant material into a long narrow heap.

yearlings Animals aged one year.

Zn Zinc

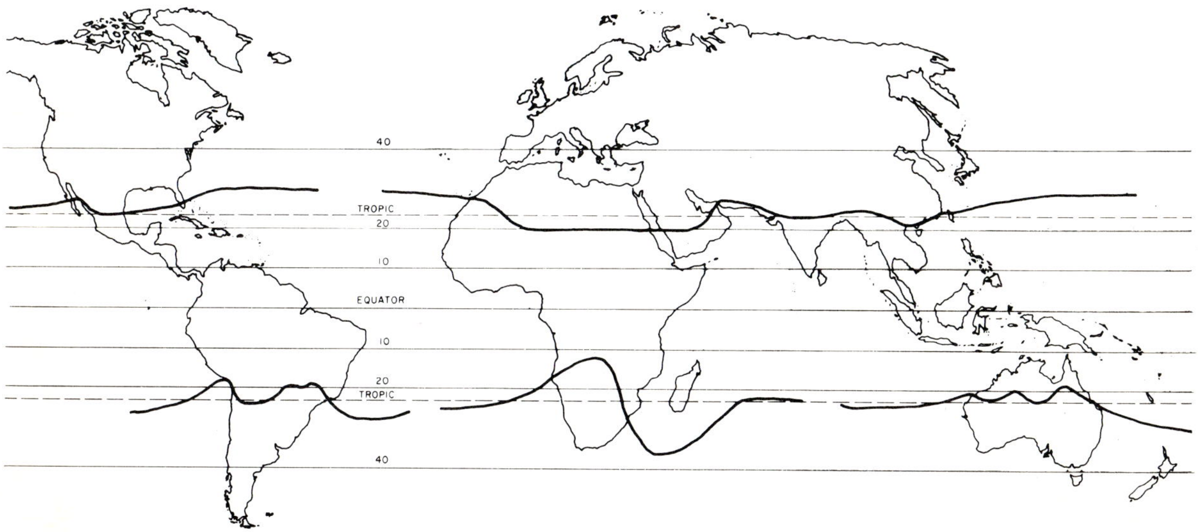

Fig. (i) Map of the world showing, in between the heavy black lines, the regions considered to be in the Tropics as defined in the Introduction.

Introduction

About half of the world's grazing animals are in the tropics, but the output of animal products from these lands is very much less than from the rest of the world. As this book will show, great scientific progress has been made in understanding the limitations to production from animals at pasture in the tropics, and in understanding how these limitations may be overcome. Developments in technology have made grassland improvement possible in most types of tropical environment, and there is now a core of livestock producers who have found grassland improvement profitable in the long term.

The first two sections introduce the main themes of the book. These are pasture and forage crop species adaptation, pasture establishment, fertiliser needs, and pasture management. Part 1 discusses the factors which control the development of natural grassland in the tropics and outlines the limitations of natural grasslands which may lead to their replacement with selected planted species or to the incorporation of additional species with them. This account is a brief one, since this book is mainly about planted pastures and students interested in the management of natural grassland are recommended to read a companion volume in this series *Range and Wildlife Management* by H. F. Heady. Part 2 summarises a philosophy of pasture improvement, which depends upon the pasture manager deciding on his production aims. He must also understand the nature of the production processes involved, recognise the factors in the environment which can be changed, and arrange the efficient use by animals of the additional better quality feed grown.

The book then discusses what plant qualities the pasture manager looks for in improved species. The book also illustrates the characteristics of many of the commercially used tropical pasture and crop plants. The suitability of plants for different climatic and soil conditions is outlined, and the principles of seed production described. Part 4 discusses pasture seed quality and the environmental factors which limit successful pasture establishment. Agronomic techniques of sowing and planting are described, both for fully-prepared seed bed situations and for establishment in natural pastures or with companion crops. The function of plant nutrients in pasture production, the fixation of atmospheric nitrogen by pasture legumes, the role of the animal in cycling nutrients back to the pasture, and fertiliser practice are discussed in Part 5.

On most farms the effort used in improving pasture growth and nutritive value will be wasted unless animals efficiently convert pasture to animal products. The final section outlines the response of the pasture (sward) to grazing and cutting, the factors which influence the optimum number of animals pastured, and the way these may be controlled on different pastures. The possibilities or options of maintaining continuity of feed supply through fodder conservation or through the development of year-round pasture grazing systems are enumerated; many choices are open to the skilled pasture manager.

In this book the tropics are considered with the subtropics. One thing that all these areas have in common is a warm environment with rain occurring during the warm season. Figure (i) shows the area north and south of the equator which is included in the 18°C isotherm for the coolest month of the year. Over the oceans this extends northwards to about 30° over the western Atlantic and southwards to 38° in the Indian Ocean south of Madagascar. Over the continents this isotherm extends north to 27° in Florida, and south to 28° in South Africa. Discussions of grassland improvement in this book will apply to lands beyond the 18°C coolest month isotherm, e.g. to southern Brazil, Paraguay and Uruguay, southern U.S.A., and to southern Queensland and northern New South Wales, since many tropical pasture species grow throughout this wide range of latitude.

Pasture plants are described by botanical names, first a genus, e.g. *Panicum*, then a species, e.g. *maximum*. These are universal in their application. Many readers of this book will know plants by common names, which vary even between districts in the same country. Some species have improved varieties selected or bred, which have superior qualities. These varieties are called cultivars (abbreviation cv.). Thus one variety of *Panicum maximum* is cv. Makueni. Grass genera are grouped into tribes, all of which form the botanical grass family. Terms which may be unfamiliar to readers are defined in the glossary on page v of the book.

Part 1 Natural grasslands of the tropics

1 Factors controlling the development of natural grassland in the tropics

The term 'grassland' is used for any plant community which includes herbaceous species. It includes forest containing a sparse understory of low growing plants and woodland with a light grass cover. It also includes savanna bearing a conspicuous grass cover but containing woody plants, true grassland where woody plants are absent and desert or steppe formations where the mixed vegetative cover is very open. Grassland is often classified according to the height of the herbaceous species; tall grassland is usually higher than 90 cm (3 ft) and short grassland less than 45 cm (18 in). It may be dominated by *perennial* species, in which vegetative buds survive the whole year, or by *annual* species which complete their life cycle during one growing season and have a seed resting stage during the dry season. It may be composed mainly of creeping species which form a dense mat, or of tussocky, bunch type species which form a distinct circular crown with spaces between the plants.

'Natural' grassland is grassland where there is an absence of sown or planted species. It is a serious mistake to imagine natural grassland as stable or even as a plant community moving towards a static, climax condition. Even in areas uninhabited by man or by domestic animals, the vegetation is constantly changing as a result of changing climate, natural plant migration and evolution, and natural fires and cyclones. The rate of change is of course greatly increased by the activities of man: introducing grazing animals, removing trees and shrubs, and burning vegetation.

An illustration of natural grassland classification is given in Table 1.1, which lists the main grasslands and grazing lands of the Australian tropics. These are first arranged according to the climate of the region in which they occur: humid tropical, sub-humid tropical etc. These are then further divided according to the type of vegetation with which they are associated; heaths, forests, woodlands etc. Some further subdivisions are then made into grazing lands with characteristic dominant herbaceous species. The subdivisions in Table 1.1 stop at very large units each covering more than a million ha (400 sq. miles). More detailed descriptions of the communities which occur within these units are needed for the classification to be really useful at the farm level. Grassland mapping has been carried out in most tropical regions, and quite advanced studies have been published for some territories.

The first requirement for the grassland manager is an understanding of the various factors affecting grassland change in his local area, and the signals which indicate whether the changes occurring are beneficial or harmful. The factors which define the local ecosystem are firstly the characteristics of the physical environment: climatic, edaphic (soil) and physiographic. Secondly there are biotic influences such as the natural fauna of the area, domestic grazing animals, firing, clearing and other modifications made by man.

(a) Climatic factors

Table 1.1 places climate as the most important of the factors which determine the type of grazing land. Generally, the wetter the climate, the denser is the tree cover. It is doubtful whether true grassland climaxes ever develop in humid areas, except in alpine regions or under some special soil conditions. Treeless areas in the humid tropics are usually caused by some form of human disturbance.

Climate has had an important influence on the evolution of the Gramineae tribes. The Andropogoneae have produced more species in the warm monsoon type climate with a wet/dry season. They have their apparent centre of distribution in Southeast Asia and include such grasses as *Andropogon, Bothriochloa, Cymbopogon, Dichanthium, Eulalia, Heteropogon, Hyparrhenia, Imperata, Saccharum, Sehima* and *Sorghum.*

The Paniceae are also a warm climate tribe; their distribution is greatest in areas of high annual rainfall. They appear to have originated in the western hemisphere. Principal genera include *Axonopus, Brachiaria, Cenchrus, Digitaria, Panicum, Paspalum, Pennisetum, Setaria* and *Urochloa.*

By contrast, the Eragrosteae are best developed in semi-arid and arid regions of the tropics; this tribe includes *Dactyloctenium, Eleusine, Eragrostis, Leptochloa* and *Tripogon.* Two other major tribes, the Agrosteae and the Festuceae, are characteristic of temperate regions, and

Grasslands	Vegetation	Grazing Lands	
		Types	Characteristic Species
Humid Tropical	Tropical heaths including wallum	Heathlands of low grazing value	Herbaceous species rare
	Tropical and sub-tropical forests	Forestlands of low grazing value (in part)	Herbaceous species rare
		Tropical tallgrass (in part)	*Themeda australis* *Heteropogon contortus*
Sub-humid Tropical	Tropical and sub-tropical woodlands	Tropical tallgrass	*Heteropogon contortus* *Themeda australis* *Sorghum plumosum* *Sorghum intrans*
	Tropical grasslands	Tropical tallgrass (bluegrass-browntop downs)	*Dichanthium fecundum* *Eulalia fulva*
Dry Tropical	Tropical grasslands	Tropical tallgrass (bluegrass downs)	*Dichanthium sericeum* *Bothriochloa erianthoides*
	Brigalow forest	Brigalow	*Acacia harpophylla* (regrowth) *Paspalidium* spp.
	Semi-arid shrub woodlands	Xerophytic midgrass (northern)	*Aristida jerichoensis* *Aristida contorta* *Aristida inaequiglumis* *Chloris acicularis* *Bothriochloa decipiens*
	Semi-arid low woodlands	Xerophytic midgrass (northern)	*Aristida pruinosa* *Aristida contorta* *Triodia pungens*
Arid	Arid tussock grasslands	Xerophytic tussock grass	*Astrebla squarrosa* *Astrebla pectinata* *Astrebla lappacea* *Iseilema membranaceum*
	Arid hummock grasslands	Xerophytic hummock grass (spinifex)	*Triodia basedowii* *Triodia pungens* *Plectrachne schinzii*
	Acacia shrublands	Acacia shrub-shortgrass	*Acacia aneura* *Acacia cambagei* *Myoporum* spp. *Eragrostis eriopoda*

many species of these tribes, e.g. *Agrostis, Festuca, Poa, Vulpia*, occur in the cooler climates often found in areas of high altitude.

Fluctuations in climate from year to year may cause radical changes in species composition of grassland. For example, the eastern margin in central Queensland of the Mitchell grass (*Astrebla*) downs referred to in Table 1.1 may become invaded and dominated by *Dichanthium sericeum* if a succession of abnormally wet years occurs. A period of below-normal rainfall causes the recession of *D. sericeum*, and its eventual replacement by *Astrebla* spp.

(b) Edaphic and physiographic factors

Soil formation is controlled by climate and by parent material. Often the association of particular pasture communities with a certain soil type is simply the result of differences in the moisture availability of that soil. The occurrence of treeless *Astrebla* downs in Queensland on deep cracking clay soils and of *Aristida* woodlands on sandy lateritic soils is probably a reflection of the more moist state of the latter. However, the association between pasture community and soil type is often useful in classification. In India the main grassland types of *Sehima/Dichanthium* and *Dichanthium/Cenchrus* are associated with black soils and with sandy loams respectively.

True grasslands sometimes develop on marshy land with poor drainage, which is too unfavourable for the growth of trees or shrubs. Examples are sedge swamps in coastal north Queensland; *Loudetia* associations in central Africa on sandy soils with poor drainage, and *Phragmites/Saccharum* grasslands in India. *Leersia hexandra* is a very valuable pasture grass in many flooded situations. Seasonally flooded grasslands are a major source of grazing in some regions, e.g. the llanos in the Orinoco basin of Venezuela, and in South Sumatra and Kalimantan in Indonesia. Similarly, very saline conditions may promote herbaceous covers rather than forest.

Slope and aspect modify the microclimate of the sward, and are responsible for some interesting differences in plant distribution. In the southern hemisphere, north facing slopes are warmer and drier than south facing slopes. The reverse is true in the northern hemisphere. Altitude exerts profound effects on climate; very high altitude equatorial regions in the north of South America and East Africa can exhibit a mainly temperate climate flora.

Table 1.1 A classification of the grasslands and grazing lands of Northern Australia. (*Source:* Moore, R. M. (1970). *Australian Grasslands*, A.N.U. Canberra.)

(c) Fire

Fire-induced grassland types are common throughout the tropical world. Many woody plants, e.g. *Terminalia* and *Combretum* spp. in Africa are killed by fire, and repeated burning leads to a herbaceous cover. The tall rhizomatous grass *Imperata cylindrica* has become the dominant species on many sloping humid and sub-humid, low-fertility lands of Southeast Asia, and Africa, as a result of repeated burning. In some cases this was associated with 'slash and burn' agriculture, in which the forest is hand-felled, burnt, and crops planted in the ashes. After a short cropping cycle of one to three years, the forest is allowed to regenerate for 3 to 20 years. A short cycle of use may eventually lead to permanent suppression of the forest, and dominance of *Imperata cylindrica*.

In central coastal and sub-coastal Queensland, a combination of partial clearing, fire and grazing has displaced *Themeda australis* and other species as the understory of the *Eucalyptus* woodlands. Established plants of *H. contortus* are more fire-resistant and seedling regeneration after firing is more efficient. Partial clearing led to the growth of more grass cover, and a consequent increase in the frequency and severity of the fires possible. Deliberate firing was practised to remove surplus old grass, for convenience of mustering stock and of hunting, and in the hope that additional green grass growth would be stimulated.

(d) Land clearing

Man removes trees and shrubs to provide himself with fuel and shelter, and to supply various industrial needs. The land may be replanted or allowed to revert to forest; alternatively it may be cleared for cultivation or to provide more grazing.

The amount of pasture available for grazing and the amount of tree foliage is related to the density of tree cover. (Tree foliage is a valuable reserve forage although it is less accessible.) Table 1.2 illustrates this principle for an *Acacia aneura* shrubland in south-western Queensland. An experiment was carried out at two sites: Boatman, 140 km south-

Table 1.2 Relationship between herbage yield, tree foliage yield and density of trees (*Acacia anuera*) at two sites in South West Queensland. (*Source:* Beale, I. F. (1973). *Tropical Grasslands* **7**:1 35.)

Site	Component	Tree Density (no/ha)		
		40	**160**	**640**
		Yield kg/ha*		
Boatman	Tree Foliage	240	810	2 330
	Herbage	2 030	1 150	180
Monamby	Tree Foliage	180	680	2 090
	Herbage	460	540	230

* 1971 data

NB 1 lb per ac = 1·12 kg per ha

east of Charleville, and Monamby, a drier site, 110 km west of Charleville (lat. 26°S). *A. aneura* trees were thinned to densities of 40, 160 and 640 trees per ha. The data in Table 1.2 are taken for a wet season several years after the initial thinning. The yield of tree foliage was greater at higher tree density. At low density bigger trees developed with more foliage, but this did not compensate for their smaller numbers. Increasing tree density from 40 to 640 trees per ha increased the yield of tree foliage about ten-fold.

Conversely, the amount of pasture available was lowest at high tree density. Very little herbage—180 to 230 kg per ha dry weight was present under the highest tree density. This increased at low tree density to over 2 000 kg per ha at Boatman, and to about 500 kg per ha at the drier Monamby site. The herbage present was mainly composed of *Aristida* spp, *Neurachne mitchelliana, Danthonia bipartita, Sida* spp. and *Trachymene* spp. It is obviously necessary to thin this shrubland if much herbage growth is to be expected. However, this leads to greater variations in bio-mass because of greater sensitivity to changes in rainfall. Reserves of *A. aneura* are required to lessen the effect of these fluctuations.

Destruction of edible trees and shrubs lowers the stock carrying capacity of some dry regions. For instance, in the Sahel and Sudanese zones of Africa browse is a most important fodder source for domestic livestock, and provides green leaf for longer periods of the year than is supplied by the low growing herbage species. Browse inaccessible to livestock may be hand-cut and fed to animals during the dry season.

(e) Grazing

The grasslands of Africa, which have evolved under grazing by a natural ruminant population, are generally of higher quality than the grasslands of Australia, which developed under very limited marsupial grazing. The increased grazing pressure caused by the introduction of domestic animals has caused very great changes in grassland communities. Some have been destroyed absolutely and others have been improved by skilled manipulation of livestock.

The botanical composition of a grassland reflects current and previous grazing pressure. For instance, when the *Dichanthium/Cenchrus/Lasiurus* grassland type in India is subjected to grazing on loose textured soils, the more palatable *Dichanthium annulatum* is lost from the sward, and *Cenchrus ciliaris* and *Lasiurus sindicus* become dominant. As grazing pressure is further increased, the sward is dominated by the low-growing stoloniferous *Cynodon dactylon* and by *Eleusine compressa.* These in turn give place to *Aristida* spp. and to the annual *Cenchrus biflorus.* Finally, only unpalatable annual weeds and bare soil remain.

This subject is treated more fully in Part 6.

(f) Other human activities

Reference has been made to the reversion of previously cultivated land to natural vegetation. This is often held at a grassland stage by grazing, burning and weeding activities. Many disturbed sites, e.g. roadsides, provide valuable and accessible natural grazing, and are maintained at early seral stages. Mining provides disturbed sites which may be colonised by plants with special tolerance of unfavourable conditions, e.g. mineral toxicity. Industrial, urban or farm pollution changes some grassland communities.

2 Limitations of natural grasslands

The viewpoint of this book is that, wherever it is economically possible, natural grassland species should be replaced with selected sown pasture species, or should have additional species grown with them.

fig. 1.1 Red Sindhi × native Lao yearling bulls grazing *Panicum maximum–Centrosema pubescens* at Na Pheng, central Laos

This statement is made in the full knowledge that, in the tropics, very little ruminant production is carried out on sown pasture. This is one reason why there is such a low output per unit area in the tropics. It is also recognised that a large proportion of tropical grazing lands is best retained in natural pasture. The returns from reseeding arid or semi-arid areas are both too hazardous and too small in relation to current costs. Some sown pasture has been developed in areas of 500 mm (20 in) annual rainfall, but usually it is only land receiving more than 600 mm annual rainfall (24 in) that can be considered. Some land is too steep, too stony or unsuited in other ways for cultivation, and intensive grassland improvement is not worthwhile.

What then are the disadvantages of natural grasslands, which make it worthwhile to replace them with sown pasture? When pastures are dominated by woody species, it is clearly an advantage if the latter are replaced by more edible plants, after stock requirements for shade and shelter have been met. The natural grasses of the shaded understory are insufficiently productive after the removal of trees, and the disturbance caused by tree removal provides a convenient seed bed for sowing improved species.

Often the natural grassland has a shorter growing season than that of improved species grown in the same area. A continuous supply of forage is of great importance in livestock production, and species can be found which have a greater tolerance for growth in extremes of climate. Natural selection may have favoured those plants which grow tallest and fastest. It may not have favoured those that have green forage available for the longest period. For example CSIRO workers at Gladstone, central Queensland (lat. 24°S) found that the natural grasslands dominated by *Heteropogon contortus* produced 90 per cent of their growth in the warmer months between mid-October and mid-April, although there was a significant rainfall in winter. A season with good autumn rainfall—240 mm in the April to June period—produced the same yield as a season with only 40 mm rainfall in this period.

The nutritive value of natural tropical grasslands is generally low. This is especially true of the more humid regions where rates of growth and of lignification are high. Sometimes good levels of nutritive value occur but these are usually only for short periods. Very low grazing pressures which provide the animals with maximum opportunity for selection are necessary to maintain good rates of animal growth. The seasonal decline in nutritive value as the pastures age is not as great for improved pasture species which have been selected for superiority in this field.

One reason for the rapid decline of nutritive value in natural grassland is the low proportion of legumes in the sward. Legumes occur in abundance throughout tropical grazing lands, but edible herbaceous legumes do not usually improve yield or nitrogen status. For example, the pastures of dominant *Heteropogon contortus* at Gladstone, referred to previously, contained a number of legumes, mainly *Indigofera enneaphylla*, *Glycine* spp. and *Zornia diphylla*. These native legumes contributed only from 0·3 to 4 per cent of the feed on offer, and presentation yields did not exceed 50 kg per ha. There are exceptions to this rule. For example, native legumes make a significant contribution to the forage supply in some grassland areas of Paraguay and the State of Rio Grande do Sul in Brazil.

Particular limitations can be found in certain grassland types. Annual grasslands tend to be unstable and often suffer from weed invasion; more drought-resistant, perennial grasses are available which provide superior protection against erosion. Selected creeping perennial grasses tolerate higher rates of stocking than many natural bunch grasses and are also more suited to the 'cut and cart' systems of grassland use found in many tropical areas. Fire-susceptible grasslands may be replaced by plants with buried crowns which tolerate burning.

An important question is how will natural grassland respond to improved soil fertility? The use of fertilisers to increase pasture growth and nutritive value (with perhaps, at the same time, the introduction of sown legumes to increase soil nitrogen status) will be less effective if natural grassland consists of species tolerant of low fertility conditions and unresponsive to improved nutrition.

The success of pasture improvement is partly due to the use of selected species from high fertility sites (e.g. floodplains) which are capable of high levels of production once the fertility of poor soil has been raised.

The basic consideration in deciding whether planted pastures are worthwhile is the level of output which can be attained from natural grasslands in relation to the costs involved. In many humid and sub-humid areas the replacement of the natural grasslands is an attractive alternative for grassland managers.

Further reading

Heady, H. F. (to be published) *Range and Wildlife Management in the Tropics.* Longman.

Moore, R. M. (1970) *Australian Grasslands.* Chap. 6–9, 16, 17. Australian National University Press, Canberra.

Rattray, J. M. (1960) *The Grass Cover of Africa.* FAO Agric. Studies No. 49 FAO: Rome.

Roseveare, G. M. (1948) *The Grasslands of Latin America.* Bull. 36 Imp. Bur. Past. Field Crops, Aberystwyth.

Shaw, N. H. and **Bisset, W. J.** (1955) Characteristics of a bunch spear grass (*Heteropogon contortus* (L.) Beauv.) pasture grazed by cattle in sub-tropical Queensland. *Aust. J. Agric. Res. 6*: 539.

Whyte, R. O. (1968) *Grasslands of the Monsoon.* Faber, London.

Whyte, R. O., Moir, T. R. G. and **Cooper, J. P.** (1959) *Grasses in Agriculture.* FAO Agric. Studies No. 42, Chap. 12. FAO: Rome.

Part 2 A philosophy of pasture improvement

1 Production objectives

The first step is to decide the aims of the grassland improvement programme for the area in question. One grassland improvement plan will usually have several results, but it is best to decide exactly what it is hoped that the changes in the management of the pasture will bring about.

(a) Animal production

Some pasture scientists dream of clothing the earth with an always green, tidy, close carpet of grassland. However, a more realistic objective is a continued, high output of animal product, and for most grassland situations, stock performance is the best way of measuring pasture value. The production of wool is best suited to short grasslands and does not need a pasture of such high nutritive value as is needed for the production of meat. Animals may be reared for sale to other pasture areas more suited to fattening. Milk production requires a high level of nutritive value and forage should always be available. Some dairy schemes in the tropics have been sited in environments better suited to meat production. Grasslands may be used primarily to provide draft power; crop production may be limited by the capacity of the available grazing to raise the buffalo, cattle, horses or donkeys used in farm operations.

In some communities animals have other uses in the social system. Animals may be used as a banking system, or play a part in marriage, family and friendship customs. In these circumstances, the capacity of the pasture to withstand high stocking rates may be more important than its capacity to produce milk or other domestic products.

fig. 2.1 $2\frac{1}{2}$ year old Hereford steers weighing 450 kg grazing at 1 beast per ha on *Macroptilium atropurpureum–Setaria anceps* pastures on well fertilised poor soils of the Queensland coastal lowlands near Bundaberg

(b) Crop Production

Some crop production systems depend upon a pasture or forage crop phase in the cropping rotation. Short-term pastures or 'leys' will give animal output but their primary purpose may be the maintenance of physical fertility of the soil, the addition of nitrogen to the soil system, or the control of pests, diseases and weeds which reduce crop production.

Table 2.1 Soil characteristics after improved pasture at Beewah, Queensland. (*Source:* Bryan, W. W. and Evans, T. R. (1971). *Aust. J. Exp. Agric. Anim. Husb.* **11**: 633.)

Soil Characteristic	Virgin Soil	Soil after 15 Years Pasture
pH	5·2	5·0
Organic carbon (per cent)	0·84	1·56
Nitrogen (per cent)	0·047	0·103
Sulphur (ppm)	70	205
Total Phosphorus (ppm)	18	160
Total Potassium (ppm)	31	202
Calcium (ppm)	30	320
Bulk density	1·55	1·00

Table 2.1 illustrates the great soil changes which may take place when a low fertility soil is sown with grass/legume mixtures, fertilised, and grazed. These data are taken from the 0–10 cm layer of a gleyed podzolic soil at Beerwah, Queensland (lat. 27°S, 1 630 mm annual rainfall). No nitrogenous fertiliser was applied, but the rhizobial symbiosis with pasture legumes (mainly

Trifolium repens) raised total soil nitrogen from 0·047 to 0·103 per cent. The content of organic carbon increased, and the 0–5 cm layer became distinctly organic, assisting the growth of the shallow rooting *T. repens*. This increased the water holding capacity of the soil, and the exchange capacity of the soil upon which mineral retention depends. It should be possible to utilise this increase in soil fertility through a short-term cropping phase.

Repeated culture of one crop often creates particular problems. Rhodesian tobacco farmers plagued with nematodes in their crops may use soil sterilants to overcome this problem; an alternative is a grass ley, using a species which suppresses nematodes. Pasture grass cultivars differ in their ability to do this, e.g. *Chloris gayana* cv. Katambora is more successful than cv. Pioneer. Winter cereal cropping in the sub-tropics may develop populations of wild oats (*Avena ludoviciana*) which may be controlled either by selective herbicides or by a pasture ley.

The association of tree crops with pastures has provided some of the most profitable opportunities for tropical pasture development. Beef production from pastures established under coconuts introduces another income source, and spreads land development and maintenance costs over two enterprises. Pastures control weed invasion and soil erosion, simplify coconut harvesting and improve soil structure. Legume-based pastures add nitrogen to the crop-pasture system.

(c) Site stabilisation

Pasture leys can also help control erosion in cropping lands. For some areas, e.g. dam catchments, water may be the most valuable product harvested from the land. A tight sod cover will modify the run-off characteristics of the catchment, reducing flood peaks and extending the period of stream flow. The life of the dam structure is reduced by the deposition of soil in the dam. This damage can be lessened by pasture cover. Stabilisation of road sides or mine workings needs a rapid growth of ground cover. High animal acceptance of pasture may be a disadvantage in these circumstances, and relatively unpalatable grasses are often the best for site stabilisation.

(d) Social amenities

Recently there has been an increasing interest in wild-life conservation and in some countries there is a need for areas outside the towns to be available for the recreation of people from the cities. This requires the setting aside of special grassland areas. In some low fertility districts these requirements can only be met if the productivity of grassland is improved; the desired population of fauna cannot be supported unless some of the limits to vegetation production are removed. This is one of the more recent challenges to grassland workers. The dramatic increase in bird numbers which has followed pasture improvement in some areas of the Queensland coastal lowlands illustrates what may be achieved if soil fertility is raised and sufficient shelter retained.

Turf on playing fields, paths and gardens must be able to resist trampling damage and have slow growth rate to minimise cutting requirements.

In designing a grassland improvement programme, the aims must first be decided on so that a sensible selection can be made of the grass species available. Long term production depends on the maintenance of the soil and water resources; this must be considered before any programme is begun.

2 Production processes

The natural processes involved in the production of vegetation in a grassland must be understood, as the success or failure of any plan for grassland improvement depends on these processes. The production of animals from grassland is the principal aim but this only involves a small fraction of the energy, water and nutrients in the total production system. This system may be considered as a series of interlocking cycles. Each cycle of energy, water or nutrients has inflows and outflows and connected storage pools of various sizes; the rates at which different processes take place determines the flow between the pools.

The primary cycle is energetic and the input is the sun's radiation as shown in Figure 2.2. In well grown tropical pastures about one per cent of the radiant energy is fixed through the photosynthesis of green plant surfaces, which converts atmospheric carbon dioxide to plant carbohydrate; this is distributed to various plant organs. Some of this carbohydrate is lost through plant respiration and the energetic needs of the plant for the maintenance of its tissues. Part of the remaining carbohydrate is ingested by domestic grazing animals and by other herbivores, whilst

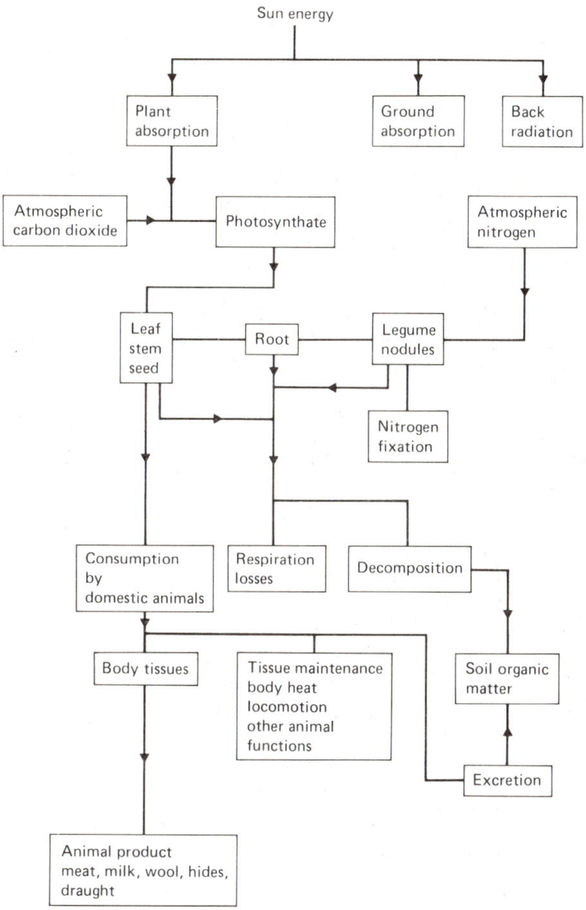

fig. 2.2

the balance is eventually broken down by the 'decomposer industry' of insects, soil fauna, fungi and bacteria.

The indigestible fraction of the food ingested by the grazing animal is excreted on the pasture. The digestible fraction contributes to the energy needs of the animal for tissue maintenance, body heat and locomotion, and to the making of the compounds which constitute animal product output—wool, meat or milk. This product output rarely represents five per cent of the energy input of the feed grazed by a ruminant at pasture. Body reserves of energy are drawn on in times of stress for survival, reproduction, lactation and wool growth.

The operation of the energy cycle depends on a satisfactory supply of moisture and of soil nutrients; these latter cycles can be modified by inputs of irrigation water or by inputs of fertiliser or of nitrogen fixed in available forms by organisms such as *Rhizobium*. Nutrient cycling is discussed in Part 5.

Increased animal output from pasture depends upon

(i) Increasing the proportion of sun energy fixed in plant tissue.
(ii) Accelerating the process by which fixed plant energy is converted to animal product.
(iii) Diverting proportionately more of the fixed plant energy to animal product.

The first two objectives involve studying the factors which limit the rates at which essential processes take place and how these factors, e.g. mineral availability, plant genetic code, may be modified. The third objective involves diverting energy and nutrients into particular channels. The proportion of photosynthate which goes to the roots and becomes unavailable to the grazing animal, the proportion of the pasture shoots which are actually grazed, the proportion of the ingested feed which is digested, the proportion of the digested feed appearing as product output are all factors which can be affected by management.

3 Changing the environment in which pastures grow

Grassland improvement involves changing the environment in which pastures grow, in order to overcome the rate-limiting factors mentioned in the last paragraph.

(a) Plant species present

The limitations of natural grassland species and the need to change the species present wherever feasible was discussed in Part 1. It should also be recognised that the introduction of selected species may also change the pasture micro-environment. The introduction of well-adapted pasture legumes adds nitrogen to the soil–plant–animal system and is a principal objective of much grassland improvement work. Improved plant productivity results from the nitrogen which legumes add to the soil. Improved animal productivity results directly from the increased protein in the legumes eaten by the animals. The use of creeping grasses, or of

fig. 2.3 Droughtmaster cows grazed on *Stylosanthes humilis* pasture in Lansdown, Australia at 0·4 beasts per ha with superphosphate application

grasses which simply grow more in that environment, provides more ground cover and modifies soil temperature and soil moisture availability. Grass roots increase soil aggregation. The removal of woody species increases the moisture available to pasture, and changes the light and temperature conditions of the sward.

Successful plant introduction uses species which are successful competitors in that environment for light, moisture, and the increased nutrient supply. Grassland managers study the competitive relations of the species in mixed swards, so that they can obtain swards with a botanical composition giving the highest animal output. This may necessitate encouraging a suitable grass/legume balance, controlling inedible or toxic species, and encouraging the plants which give the highest yield when ingested by the animals.

(b) Soil conditions

Fertiliser use on pasture to remove mineral deficiencies limiting pasture growth or nutritive value is discussed in Part 5. Other forms of amelioration, e.g. the use of gypsum to improve the physical structure of solodised solonetz soils may improve the soil conditions for pasture growth.

(c) Moisture availability

The use of irrigation on pastures is mainly confined to areas producing higher priced products such as whole milk, stud animals, or high protein forage crops. Some inexpensive water diversion schemes are in operation which collect run-off water from hard ridges and spread it over more gently sloping, absorbent soils. This provides special pasture areas with an extended growing season. Water retention on seasonally flooded areas can also extend the period of forage growth. Terracing or contour ridging reduces run-off. Cultivation can be used to restrict transpiration losses in one season in order to store moisture for forage crop use in a later period of moisture scarcity.

(d) Temperature, wind and light

Only minor modifications to the temperature and wind situations are possible. The type of pasture canopy maintained and the occurrence of trees and hedges have some influence. It is not economic to artificially increase the light supply to pasture, but the type of pasture canopy grown and the way it is grazed can greatly increase the efficiency of light use.

(e) Biotic environment

Obviously the number and kind of domestic animals grazed and the measures adopted to control the fauna population are features of the pasture environment.

4 Efficient use of pasture

Efforts made to improve the pasture species grown and the environment in which they grow are worthless unless the additional better quality forage produced is converted to animal product.

Probably the most important factor deciding animal output is the amount of green pasture eaten per unit area. Young, green pasture is higher in nutritive value and is more readily accepted by animals than older material. Often we do not appreciate the high rate at which the leaves and stems age. Leaves are lost from the plant by animal trampling, by physical erosion, and by microbial breakdown. The pasture manager seeks to divert as much of this material through the jaws and into the belly of the grazing animal as is possible. One result of high grazing pressure is that more of the photosynthate remains in the leaves and stems, and less is distributed to the roots. A proper balance needs to be maintained here. Root development is necessary for the full exploration of the soil mass upon which plant moisture and nutrient uptake depend, and for plant survival in periods of environmental stress. However, too much energy accumulated below ground, and inaccessible to the grazing animal, is wasteful. If 70 per cent of the pasture tops are eaten we have a much more efficient production system than if only 30 per cent are consumed. Efficient consumption depends first of all on having enough animals to eat the pasture. If domestic animals are not used, then the animal population of the 'decomposer industry' automatically builds up instead: for instance, the population density of small animals, e.g. arthropods in the top 6 cm of the soil has been found to be low when a large number of sheep are grazed per unit area of pasture. These key management factors of stocking rate, and the continuity of pasture supply needed to maintain high stocking rates, are discussed in Part 6.

Pasture must be of a high nutritive value for effective conversion to animal product. Tropical pastures are generally of lower nutritive value than temperate pastures, and this question is receiving a lot of attention from pasture breeders. The higher growth rates of tropical pastures permits higher stocking rates, but individual animal performance, which reflects pasture quality, is often lower. Long term gains of 0·5 to 0·6 kg liveweight per day can be obtained in most situations, but gains of the order of 1·0 kg per day usually only occur for short periods. Consequently the grassland manager has to watch constantly the quality of the forage available and control it to maintain animal performance. The factors involved in nutritive value are discussed at the beginning of Part 3. The genetic capacity of the animal to convert forage to useful product and the health of the animals are further limitations. However these questions have often received too much attention in districts when stock are undernourished and this has caused great financial loss to livestock producers and governments. Investments in stud animals and vaccination programmes cannot provide worthwhile returns where the level of forage supply is still the first limitation to animal output.

The cycling of nutrients from the animal to the soil-pasture system and then through the animal again, allows nutrients to be used several times. Almost 90 per cent of the nitrogen in the pasture ingested by the animal is voided on the pasture as urine and dung. Some of this is lost by denitrification to the atmosphere, or by leaching, or is concentrated in great amounts on areas where stock gather together. However, the interruption of this cycle can have very serious consequences on pasture production. Many tropical small-scale producers use a 'cut and carry' system of production, in which pasture is cut and removed to an animal feeding place. Some large scale 'green chop' dairy operations are also in operation. If the dung, urine and reject forage are not carefully collected and returned to the pasture, the reduction in pasture soil fertility is very rapid. Potassium deficiency can appear very quickly in fields cut for hay and silage, and this increases the fertiliser needs of the pasture. This cycle is discussed in Part 5.

5 Management options

The grassland manager has many paths leading to increased animal output from improved pasture. Some paths are much better than others for particular farm situations. The farmer will make more sensible management decisions if he understands the limitations of the environment and the chain of processes leading to product output. Market needs, farm costs and product prices are constantly changing, altering the profitability of each farm improvement practice. This can only be assessed if the manager has good information available to him on both the benefits of a practice and the cost of the practice. This assessment of costs and returns needs to take into account more than the short-term cash effects; the grassland manager is always concerned to avoid damage to the basic environmental resources as this is what his whole operation depends on. He is also concerned about the risks involved; what is expected from successful grassland improvement, and will any hazards to animal survival be changed?

Concentrating on the weakest links in the production

fig. 2.4 Three-quarter Brahmin cows grazing *Centrosema pubescens–Setaria anceps* pastures at Na Pheng, central Laos

process will always yield the highest returns for the effort and money invested. For one producer this may be the forage available to cows during late pregnancy, which controls calving losses, calf growth and subsequent conception. For another producer, slow growth of weaners in the first six months after weaning may require special attention. A crop producer who depends on buffalo for early ploughing may have his effective crop area limited by buffalo body condition at the beginning of the wet season.

The definition of these problems and an understanding of how these production limitations may be overcome will help the grassland operator decide which management options are the best in his own farm situation. The principles involved in choosing well adapted pasture species, and in establishing, fertilising and grazing them are discussed in the next four chapters.

Further reading

Humphreys, L. R. (1966) Pasture defoliation practice—a review. *J. Aust. Inst. Agric. Sci. 32:* 93.

Hutchinson, K. J. and **King, K. L.** (1970). Sheep numbers and soil arthropods. *Search 1:* 42.

Hutton, E. M. (1970) Tropical Pastures. *Adv. Agron. 22:* 1.

Shaw, W. M. and **Bryan, W. W.** (Eds.) (1976) *Tropical Pasture Research.* CAB, Hurley.

Part 3 New and improved pasture plants

1 Desirable plant qualities

The suitability of improved pasture plants for the areas in which they are to be grown depends on the production objectives discussed in the last chapter. The environment in which the pastures are grown must also be considered. The aims of pasture use (e.g. the type of animal production, the importance of pasture-crop integration or the necessity for watershed stability) will decide the importance of various pasture qualities (e.g. nutritive value, ease of eradication, or ground cover). The characteristics of the local district will decide the relative importance of such factors as drought or flood tolerance, or soil fertility building. A useful discussion of climatic zones applicable to tropical species will be found in Reid (1973).

The need for plants to have versatility of use and wide environmental adaptation is an issue on which scientists have varied opinions. According to the results of some evaluation programmes, high performance in different environments under a wide range of site management conditions is of the greatest importance. It is argued that varieties with these characteristics will rarely fail in the hands of the farmer, even when abnormal seasonal conditions occur or grazing management alters. Some pasture varieties, e.g. *Macroptilium atropurpureum* cv. Siratro and *Desmodium intortum* cv. Greenleaf have been released with a wide genetic base, and some adaptation to local districts may be expected as genetic shift occurs.

The other approach is to select plants which only perform well in specific pasture conditions and have limited uses. The seed of such a variety has a much smaller market, and can only be successful in a very advanced farming community. A narrow genetic base also leaves a variety more susceptible to the evolution or introduction of new pests and diseases. Today, in most tropical countries, producers look for pasture plants which perform well over a wide range of farm conditions.

Pasture varieties are usually judged in terms of their nutritive value, growth, ease of establishment, and persistence. Those varieties to be used for special purposes must be judged in other ways as mentioned later in the chapter.

fig. 3.1 A *Desmodium intortum* dominant pasture with *Digitaria decumbens* at Mt. Cotton, south-eastern Queensland.

(a) Nutritive value

The feeding value of tropical pastures is given first consideration in this discussion, since it is the characteristic most needing improvement. The efficiency of conversion of tropical pastures to animal product is lower than that of most improved temperate pastures. This makes really high individual animal performance harder to attain. On the other hand, the superior growth rate of tropical pastures compensates for this, since higher stocking rates are feasible. Nutritive value is assessed in terms of energy availability, and the content of protein, minerals and vitamins, and the absence of toxins.

(i) Energy

Animal output is first determined by the amount of forage eaten and the proportion digested. There should be no deficiencies of minerals or protein and if these are in adequate supply then animal performance is controlled by the energy level of the forage. The amount of pasture an animal eats varies considerably between species and between cultivars. Species giving a dense sward structure provide a larger intake per bite than loose, trailing species.

Comparisons of feed intake in the field are difficult to make. Much of our information comes from animal house experiments with chopped feeds which do not always provide sound guides to the field behaviour of grazing animals, although they are applicable to 'cut-and-carry' feed systems.

Table 3.1 compares the voluntary intake of four *Panicum* varieties grown at Gatton, Queensland (lat. 28°S) under conditions of good nitrogen supply. Dry matter intake of good quality pasture is usually about 2·5 to 3 per cent of body weight, but a more precise comparison is made by referring it to the 'metabolic size' of the animal, which is expressed as kg body weight $(W)^{0.75}$. Dry matter intake was measured from series of pastures sampled after various intervals of regrowth after cutting.

In the first regrowth series, intake of variety CPI 13372 decreased from 57·6 units 28 days after cutting to 36·4 units 91 days after cutting. For the same series, intake of Coloniao variety decreased from 69·2 units 28 days after cutting to 52·1 units 91 days after cutting. In the second regrowth series later in the season the values for voluntary intake were generally lower, but the general decrease in intake as the pasture aged was also less.

This table shows quite large differences in voluntary intake between pasture varieties: for the average of both series, CPI 13372 had the lowest value of 50·0 units and Hamil the highest value of 62·8 units. Varieties also varied in the rate at which voluntary intake decreased as the pastures aged. These are both important selection criteria.

Table 3.2 shows the dry matter digestibility of the same pastures. Apparent digestibility is calculated as

$$\frac{\text{feed consumed} - \text{faecal output}}{\text{feed consumed}} \times 100$$

Digestibility can be calculated for particular constituents of the feed, and may be also expressed in terms of energy, dry matter, or organic matter. Dry matter digestibility is usually 2 to 3 per cent lower than organic matter digestibility, and the two are very highly correlated. For the same regrowth series, digestibility of Coloniao decreased from 62·2 to 50·2 per cent. Values later in the season were lower, and decreased more slowly with age. It is usual for the rate of decrease in digestibility with age to be greater for pastures which began with a high nutritive value and are growing rapidly.

Mean digestibility varied between the varieties from 51·0 per cent for CPI 13372 to 54·8 per cent for Coloniao. This is a much smaller difference than the difference in intake previously discussed, and points to the need for information on the voluntary intake of the selected pasture plants under test. Digestibility can be accurately and easily estimated in the laboratory by means of the *in vitro* technique, in which rumen liquor collected from an animal is used to digest small samples of pasture held at constant

Table 3.1 Dry matter intake (g/kg 0.75 weight) of four varieties of *Panicum* sp. (*Source*: Minson, D. J. (1971). *Aust. J. Exp. Agric. Anim. Husb.* **11**. 18.)

Harvest Time		CPI 13372	Green panic	Coloniao	Hamil
First regrowth series:					
December	28 days	57·6	59·2	69·2	68·8
January	63 days	43·5	47·6	51·6	57·1
February	91 days	36·4	42·3	52·1	62·2
Second regrowth series:					
April	42 days	55·4	53·9	65·1	60·8
May	70 days	57·5	48·7	65·0	64·2
June	98 days	49·9	46·2	61·8	63·6
Mean of both series:		50.0	51·3	60·8	62·8

Table 3.2 Dry matter digestibility (%) of four varieties of *Panicum* sp. (*Source:* Minson, D. J. (1971). *Aust. J. Exp. Agric. Anim. Husb.* **11**:18.)

Harvest Time		CPI 13372	Green panic	Coloniao	Hamil
First regrowth series:					
December	28 days	59·3	62·0	62·2	59·5
January	63 days	50·3	47·4	53·4	50·3
February	91 days	43·2	47·7	50·2	48·2
Second regrowth series:					
April	42 days	54·4	52·0	55·8	54·4
May	70 days	50·0	49·0	54·0	51·0
June	98 days	48·9	47·6	53·3	51·1
Mean of both series:		51·0	51·0	54·8	52·4

temperature for a standard period. The correlation, or agreement, between values for digestibility and values for intake is not as high in tropical pastures as we might hope, although the rate of digestion under *in vitro* conditions shows quite a good correlation with intake. Comparing Tables 3.1 and 3.2, it is seen that the two varieties with higher digestibility, Hamil and Coloniao, also had higher intake values than the other two varieties. However, the digestibility of Coloniao was superior to that of Hamil, although sometimes Hamil had a higher intake value.

Some plant organs are more digestible than others and some structures, e.g. thorns, may limit intake. Young lamina is superior to leaf sheath and to fibrous stem; flowers are more variable in quality. Consequently, many people emphasise leafiness as a desirable quality, and regard grasses with a long season of inflorescence production as inferior. The changing proportions of leaf, leaf sheath, stem and inflorescence account for some overall differences in cultivar nutritive value. For instance, Hamil and Coloniao varieties did not flower in the experiment reported in Table 3.1. However, young stem material is highly digestible, and may be preferred to senescent lamina. There are no simple, reliable, ways of deciding, by eye, the energy value of plants. Measurements of leaf tensile strength and leaf rigidity are promising indicators. Usually varieties which always bear accessible, young active growth give good performance and, within a particular

fig. 3.2 Open panicle of *Panicum maximum* var. *trichoglume*

species, it is best to select the leafier cultivars.

The *in vitro* estimate of digestibility is most important in laboratory comparisons of nutritive value. For many decades, laboratory assessment of feeding value has mainly depended on 'proximal' analysis. This estimates soluble carbohydrate, fat (extracted by ether), crude fibre, crude protein and minerals. The systems of Starch Equivalent (SE) or Total Digestible Nutrients (TDN) derived from these estimates are used in assessing fodder quality in temperate countries. However, these systems are not necessarily well related to production from tropical fodders. The fibre fraction is of particular importance for digestibility and intake, but because it contains so many diverse types of compounds its value is difficult to measure. Some workers have related digestibility to the content of hemicellulose and cellulose present and the extent to which these are lignified. Tropical pastures are often grown under conditions where some accumulation of standing feed is necessary. In these circumstances, it is an advantage to have plants where the structural materials (e.g. the polysaccharides of the cell wall) are highly digestible. As an example of this, the digestibility of total cell wall polysaccharides is higher in *Digitaria decumbens* than in *Setaria anceps*.

fig. 3.3 Spike-like panicles of *Setaria anceps*

Considerable attention has also been paid to soluble carbohydrate content, which is a good measure of the immediate energy content of the forage. About 60 to 70 per cent of the animal's energy requirements are absorbed through the rumen wall as volatile fatty acids (V.F.A.). Thus energy retention in the animal must be favoured by a high level of V.F.A. production. The principal V.F.A. for meat and milk production are acetic, propionic and butyric acids and their relative values have been assessed. However, so far conclusive evidence has not been forthcoming that V.F.A. analysis can be used to select pasture varieties best suited to a particular type of animal production.

(ii) Protein

Correction of protein deficiency in tropical pastures once seemed to be the major challenge to scientists. In field practice inadequate protein supply during part of the year is still a major limiting factor to stock production. However, for most tropical environments techniques are now available to overcome this problem, whilst carbohydrate quality remains the major limitation. Protein needs may be met by feeding legumes of high protein content, by supplementary feeding of high nitrogen concentrates such as urea, by selecting grasses with better nitrogen response characteristics, by the application of nitrogen fertiliser, or by allowing animals to selectively graze the parts of the plant which are high in nitrogen.

Table 3.3 summarises the results of an experiment in which sheep were fed different diets. If mature *Digitaria decumbens* of 3·6 per cent crude protein content were given as feed, daily intake by sheep was only 607 g per day. As the legume *Medicago sativa* was added to the ration and protein deficiency was corrected, the intake of *Digitaria decumbens* increased. When the mixed diet contained 7·2 per cent crude protein, sheep averaged 844 g grass plus 200 g legume intake per day. At high levels of legume feeding

Table 3.3 Intake and crude protein content of mature *Digitaria decumbens* and *Medicago sativa* diets. (*Source:* Minson, D. J. and Milford, R. (1967). *Aust. J. Exp. Agric. Anim. Husb.* **7**:546.)

Diet	Dry matter Intake (g/day)		Crude Protein (per cent)
	D. decumbens	*M. sativa*	
D. decumbens* only *ad lib	607	0	3·6
***D. decumbens* + 100 g legume**	783	100	5·7
***D. decumbens* + 200 g legume**	844	200	7·2
***D. decumbens* + 400 g legume**	682	400	10·3
Legume *ad lib*	0	1 530	22·5

NB 1 lb = 453·6 g

the sheep ate less grass. This experiment emphasises the importance of finding new and improved pasture legumes to correct grass protein deficiency and so increase both the efficiency of grass use and the total energy intake.

The superiority of legumes in this respect is indicated in Figure 3.4, which shows the comparative intake of digestible dry matter (i.e. intake × dry matter digestibility) for some tropical legumes and a group of tropical grasses. These data were obtained from pen feeding experiments at Gatton, Queensland (lat. 28°S) and most of the legume values were obtained during the dry and frosty winter period, when feed quality is very important.

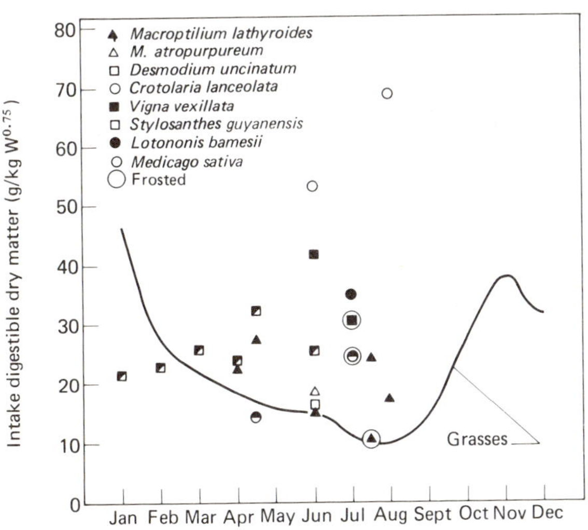

fig. 3.4 Comparison between the intake of digestible dry matter of eight pasture legumes and the monthly means for some tropical grasses. (*Source:* Milford, R. (1967). *Aust. J. Exp. Agric. Anim. Husb.* **7**:540)

Protein quality is not of major significance for ruminant animals at pasture, since most of the necessary amino-acids can be synthesised by rumen organisms if the nitrogen supply is adequate. Hence nitrogen content (or crude protein content = 6·25 × % nitrogen) is the primary selection criterion. Percentage nitrogen of the feed actually ingested should be at least 1·1 for maintenance. 1·6 per cent nitrogen is generally satisfactory for beef production, whilst dairy cows require about 1·9 per cent nitrogen for maximum forage intake.

(iii) Minerals and vitamins

Keeping dietary levels of minerals adequate in the feed is usually a local problem, sensitive to soil type and rainfall pattern. Deficiencies may be overcome indirectly by fertilising the pastures, and this is recommended where mineral deficiencies are also limiting pasture growth. Where the deficiency appears in the animals rather than the pasture, direct feeding of mineral licks or supplementation via drinking water is recommended.

However, it should also be recognised that pasture plants vary in their mineral content, and animal disorders are less common when animals graze particular species

Table 3.4 Sodium content of some tropical pasture species. (*Source:* Playne, M. J. (1970). *Aust. J. Exp. Agric. Anim. Husb.* **10**:32).

Plant	Sodium Content (per cent DM) Range	Mean
Chloris gayana		
cv Pioneer	1·34–1·79	1·61
C. gayana		
cv Callide	0·18–0·61	0·44
Panicum maximum	0·83–1·43	1·00
Urochloa mosambicensis	0·04–0·88	0·27
Cenchrus ciliaris	0·02–0·31	0·09
Stylosanthes humilis	0·007–0·154	0·06
Macroptilium atropurpureum	0·015–0·042	0·02
Heteropogon contortus	0·005–0·041	0·02

fig. 3.5 Digitate panicles of *Chloris gayana*

whose mineral concentrations fall within the desired limits. For example, work by the Queensland Department of Primary Industries on copper deficient coastal lowlands showed that the copper levels in *Trifolium repens* and *Paspalum dilatatum* were much higher than in *Axonopus affinis*, even where no copper fertiliser was applied. Similarly, CSIRO work at Lansdown (lat 20°S) found very wide variation in sodium content species and cultivar.

Normally it is considered that 0·05 to 0·15 per cent sodium in the diet is satisfactory for most classes of stock. Table 3.4 indicates that on the high sodium solodic soil where the plants were grown, *Stylosanthes humilis, Macroptilium atropurpureum* cv. Siratro and *Heteropogon contortus* often had very low sodium content. On the other hand, plants like *Chloris gayana* contained high sodium levels, cv. Pioneer accumulating more than cv. Callide.

Phosphorus deficiency is very common and some improved species such as *Stylosanthes humilis* have rather low phosphorus contents, even when fertilised with phosphatic fertiliser. Potassium can lead to alterations in the sodium and magnesium balance in forage plants, and thus lead to secondary deficiencies. Sulphur is important for the synthesis of some aminoacids such as cystine and methionine which affect the rumen organisms.

Micro-element deficiencies such as cobalt, iodine and copper should be considered; the level of copper supply decreases when molybdenum availability increases. Excess of minerals, for instance of lead or selenium may occur, and plants differ greatly in their tendency to accumulate selenium.

Vitamin levels are usually adequate in grazed pasture. Housed animals receiving dry feeds only may respond to vitamin A, but the deficiency is not as widespread as was once believed.

(iv) Toxic substances

New pasture introductions need to be screened for their possible content of harmful or toxic substances. For example, an otherwise very promising legume *Indigofera spicata* contains a dangerous hepatotoxin, indospicine. *Leucaena leucocephala* has a high content of mimosine, which depresses cell division. Monogastrics such as chickens lack a system of detoxifying this compound, and some damage may result if *L. leucocephala* forms a high proportion of the diet. This problem is not severe in ruminants producing meat or milk.

Fortunately animals grazing tropical legumes are almost entirely free of bloat; this may be associated with the relatively high tannin content, which is high relative to temperate legumes such as *Medicago sativa.* Levels of oestrogenic substances in tropical legumes do not cause concern.

Toxicity problems occasionally arise with grasses. The high oxalic acid content of *Setaria anceps* has killed cattle.

This phenomenon seems to be related to the abnormally high potassium requirement of the cultivars of this species which have become available. This high cation uptake needs to be balanced by anions and the plant uses oxalic acid to maintain ionic balance. One CSIRO experiment showed that the 'critical' potassium concentration—that is, the potassium concentration in plant tops needed for 90 per cent maximum growth—was 1·6 per cent in cv. Narok, 1·8 per cent in cv. Nandi and 3·1 per cent in cv. Kazungula. Oxalic acid levels showed a similar pattern between varieties.

Further selection work is needed to identify plants of *Setaria anceps* which balance cations with harmless anions, or which have a lower potassium requirement.

Tainting compounds which reduce the quality of milk or meat also require screening.

fig. 3.6 Pods present on the tall shrub legume *Leucaena leucocephala*, grown with *S. humilis* and *Heteropogon contortus* at Rodds Bay, Central Queensland

(b) Growth

(i) Total yield

Yields from tropical pastures are much higher than from temperate pastures. The best yields reported from well watered, well fertilised, improved species are of the order of 40 000 to 50 000 kg dry matter per ha in the tropics, and 15 000 to 20 000 kg per ha in the temperate regions. This is due to many factors. Temperature conditions in the tropics are more favourable for growth, and greater amounts of radiation are received. Tropical grasses are able to respond to higher intensities of radiation than temperate grasses, and have a different photosynthetic pathway (called the C_4 dicarboxylic acid pathway) from the C_3 phospho-glyceric acid pathway of dicotyledonous species and the temperate grasses. Most plants use up carbohydrate in respiration, losing carbon dioxide throughout the day and night; however tropical grasses do not lose carbon dioxide during daylight.

Total yield must be considered since superior performance both creates a larger energy pool for conversion to animal product, and ensures a competitive advantage for the variety in question, which helps it to persist. Plant growth is the product of the capacity of the photosynthetic system (the extent of its green surfaces, L) and the efficiency of its photosynthetic system (the balance between photosynthesis and respiration per unit green surface area, or net assimilation rate, Ea). Thus

$$\text{Growth rate} = L \times Ea$$

Assimilation rate is often different for different varieties, e.g. in *Cenchrus ciliaris* cultivars, but differences in the amount of leaf surface area are more important in selecting for high yield. These depend on differences in the rate of leaf appearance, expansion, and senescence.

(ii) Distribution of assimilate

Assimilate produced in the leaves of plants may be distributed to the meristematic centres of leaf production, to stem tissue, to developing inflorescences or to underground plant organs. The balance of distribution can affect the usefulness of pasture plants for different uses. For instance, a higher proportion of assimilate in *Cenchrus ciliaris* is transported to the roots and underground rhizomes compared with *Panicum maximum* var. *trichoglume* cv. Petrie. *C. ciliaris* is more drought resistant and more tolerant of heavy grazing than *P. maximum.* On the other hand, *P. maximum* has more of its assimilate accessible to grazing animals, and is a better ley species, easier to eradicate and giving higher subsequent crop yields than *C. ciliaris.*

Self-regenerating annual legumes such as *Stylosanthes humilis* transport much of the carbohydrate produced in the plant green surfaces to the seeds, and seed yields in excess of 1 000 kg per ha can occur in well grown swards. This is desirable, since the seeds form a high protein diet for stock during the dry season and ensure the plant's sub-

fig. 3.7 University of Udayana personnel weighing cattle on pastures under coconuts at Sangiang, west Bali

sequent regeneration. On the other hand, perennial pasture grasses which produce flowers over a long period produce a great deal of structural tissue of poor nutritive value, and seed is difficult to recover. A more desirable characteristic is a well defined grass flowering response to shortening days which delays inflorescence production until late in the growing season.

(iii) Seasonal distribution of growth

Animal requirements for forage are usually more constant during the year than pasture growth rates. In these circumstances, pasture cultivars which maintain good year-round growth rates, with less fluctuation between peak and minimum values, have a distinct appeal. It is argued that there is little virtue in growing a great mound of pasture which is more than the needs of the animals during the peak growing season. This is particularly true if environmental or management conditions do not favour its conservation for later use. Thus a capacity to grow well over a wide range of environmental conditions is very useful. This may depend upon a deep rooting habit which exploits deep soil moisture reserves during a dry season. It may involve capacity to grow under cool conditions; for example, most cultivars of *Setaria anceps* and *Pennisetum clandestinum* make more cool season growth than *Digitaria decumbens.*

Although we may try to keep the variation between warm and cool season growth to a minimum, we should not overlook the fact that the objective is more cool season forage. For example, CSIRO work with *Setaria anceps* found that some frost-tolerant introductions from high altitude Kenya had a higher *ratio* of winter to summer yield in Queensland than other varieties, e.g. Kazungula. However, the *amount* of winter yield that each variety produced agreed well with the summer yield of that variety. Vigorous varieties such as Kazungula which had a lower ratio of winter to summer yield produced good winter yields.

(iv) Responsiveness to nutrients

Pasture species highly responsive to nutrients are needed for production improvement situations using high fertility land or fertiliser inputs. For instance, work at Serere in Uganda has shown that *Panicum maximum* and *Chloris gayana* are best used under conditions of good nitrogen availability, whilst *Hyparrhenia rufa* is clearly better adapted to poor soils of low organic matter status.

Some pasture improvement plans rely on plants which are most efficient in extracting nutrients from low fertility soils. In comparing the yield performance of different cultivars one needs to know the fertility conditions under which the comparison is made.

(v) Legume nitrogen fixation

When choosing pasture legumes, the most important consideration is the amount of nitrogen fixed by the rhizobial symbiosis. Nitrogen fixation has been included in this section on growth, since it depends mainly upon the rate of legume growth.

Farmers often ask whether one legume is a better nitrogen fixer than another. If both legumes are well nodulated, the answer is usually that the legume which grows the most fixes the most nitrogen. For instance, in a moist environment which favoured the growth of *Stylosanthes guyanensis*, this plant might fix more nitrogen than *S. humilis*. In a drier environment to which *S. humilis*

fig. 3.8 Drooping panicles of *Hyparrhenia rufa*

was better adapted than *S. guyanensis, S. humilis* might fix more nitrogen than *S. guyanensis*.

Basically, nitrogen fixation depends upon the energy conversion of surplus assimilate. The amount of assimilate available depends upon the growth rate. Some of the surplus is transported from the legume shoots to the nodules on the roots. Usually between 20 and 40 kg nitrogen are fixed for every 1 000 kg dry matter of shoots grown.

Genetic differences in the efficiency of the rhizobial symbiosis also occurs and in some cases decides which variety is chosen. For instance, CSIRO work has shown considerable difference between varieties of *Glycine wightii* in their capacity to nodulate, and between rhizobial strains in the effectiveness of their symbiosis; cv. Cooper gave poor performance relative to other cultivars such as cv. Tinaroo. Total nodule weight per plant seems a good guide to selection.

(c) Ease of establishment

(i) Reliability of establishment

Plants which are easy and reliable to establish are accepted

fig. 3.9 A well grown seed crop of *Chloris gayana*

more easily than plants with poor seedling vigour or with poor rooting capacity as vegetative material. *Panicum coloratum* var. *makarikariense* is one pasture plant with many desirable characteristics, whose use has been more limited than might be expected from its performance as well established pasture. Table 3.5 gives some comparative data with *Panicum maximum* var. *trichoglume* cv. Petrie for the establishment year and for the subsequent growing season at Toowoomba, Queensland (lat. 28°S). In the first season, yield of *P. coloratum* was 24 g relative to 58 g for *P. maximum*; however in the second year *P. coloratum* varieties greatly outyielded *P. maximum*. Slow seedling growth of *P. coloratum* was associated with delayed

fig. 3.10 *Panicum coloratum* var. *makarikariense* growing in Kenya. The stolons are forming roots at the nodes

Table 3.5 Yield and tillering of *Panicum* cultivars in successive seasons. (*Source:* Lloyd, D. L. (1971). *Aust. J. Exp. Agric. Anim. Husb.* **11**:525.)

Character	Cultivar		
	P. coloratum cv Pollock	*P. coloratum* cv Bambatsi	*P. maximum* cv Petrie
Yield (g/plant):			
1st season 12 weeks	24	24	58
2nd season 17 weeks	396	229	107
Tiller (no./plant):			
1st season 12 weeks	24	18	45
2nd season 17 weeks	155	95	43

NB 1 lb = 453·6 g

tillering, especially in the first 16 to 25 days after emergence. Many farmers are unwilling to give the care and attention to seed bed preparation and to first year management of the sward which a species of this type demands.

Factors affecting reliability of establishment are discussed in Part 4. Many problems are avoided by using large seeded varieties. In the case of legumes, varieties which do not benefit from inoculation with *Rhizobium* are simpler to establish. Plants easy to establish also spread more rapidly to adjacent areas.

(ii) Seed production characteristics

Many tropical pasture selection programmes have placed most importance upon yield, nutritive value and persistence, and have given little attention to seed production characteristics. Some improved cultivars produce flowers over a long period and these ripen unevenly and the seeds shatter, so that little seed is available for harvest at any one time. This leads to shortages of supply and inevitably higher prices.

Large differences in seed production potential exist among released varieties; for instance, in Queensland a single commercial harvest from *Setaria anceps* cv. Nandi might yield 50 kg seed per ha, compared with 200 kg seed per ha from *Paspalum plicatulum* cv. Rodds Bay. Within a single species there is ample scope for improving seed production characteristics. Table 3.6 shows the variation available from selection within *S. anceps* cv. Nandi in Kenya. Single plant clones were established 1 m apart. Head characteristics were similar but inflorescence density varied from 42 per m^2 for the mean of the unselected population to 148 per m^2 for the mean of the three best seed producing clones. The quality of seed also varied between clones, and this factor was often inherited.

Table 3.6 Genetic variation in seed yield of *Setaria anceps* cv. Nandi clones. (*Source:* Boonman, J. G. and van Wijk, A. J. P. (1973). *Neth. J. Agric. Sci.* **21**:12.)

Seeding Characteristics	Mean of 18 Clones	Mean of 3 selected Clones
Inflorescence density (no./m^2)	72	148
Inflorescence length (cm)	19·7	17·8
Seed (no./cm inflorescence)	2·25	1·96
Seed weight (mg)	0·52	0·54
Seed yield (kg/ha)	16·6	27·9
Per cent pure live seed	27·1	49·8

NB 1 lb per ac = 1.12 kg per ha

Some plants in commercial use e.g. *Digitaria decumbens* do not produce seed, and improvement in planting techniques has made large scale vegetative propagation

fig. 3.11 Racemose panicles of *Paspalum plicatulum*

quite feasible. Many small farmers prefer to establish pastures from roots or cuttings as they can easily build up their own supplies of planting material and do not require safe seed storage.

(d) Persistence

Pasture plant persistence depends on how successful the plant is in adapting to the environment and also on the plant's degree of perenniality. 'Persistence' is sometimes also loosely applied to annuals such as *Stylosanthes humilis* to indicate their success in self-regeneration from year to year. Obviously pastures which last and do not require resowing return more on the original improvement investment than pastures which have to be frequently reploughed and reestablished.

Table 3.7 Effect of severity of defoliation on *Desmodium intortum* lines. (*Source:* Imrie, B. C. (1971). *Aust. J. Exp. Agric. Anim. Husb.* **11**:521.)

Character	Defoliation	CPI 23189	Mean of 6 Lines
Leaf weight (g)	Heavy	22·1	13·9
	Lenient	22·8	23·0
Stem weight (g)	Heavy	11·9	8·8
	Lenient	14·5	16·8
Leaf number	Heavy	595	440
	Lenient	441	519
Stem number	Heavy	144	105
	Lenient	119	130

Persistence may be measured quite simply as the way a plant continues to contribute to yield. It may also be assessed in terms of the percentage of ground covered by a variety. It is usually determined by the reaction of the plant to grazing, to environmental extremes, to insect or disease attack, or to its companion plants.

(i) Reaction to grazing

The reaction of pasture plants to defoliation and to trampling is discussed in Part 6. The most resistant plants are usually those with some buds inaccessible to the jaws of the grazing animal—plants with rhizomatous, stoloniferous or tall shrub habit—and with a capacity to regrow quickly after defoliation. This character can be selected for successfully in many species.

Table 3.7 shows some variation available within *Desmodium intortum*, a species which is not particularly tolerant of heavy grazing. Six lines of *D. intortum* were subjected to heavy or lenient defoliation. It will be seen from Table 3.7 that heavy defoliation substantially decreased the mean leaf weight of the six lines from 23·0 to 13·9 g, and the mean stem weight from 16·8 to 8·8 g. On the other hand, the leaf weight of CPI 23189, a line from the Philippines and one of the original components of cv. Greenleaf, was unaffected by severity of defoliation. This was associated with a higher density of axillary buds close to the plant crown, and smaller, more numerous leaves.

fig. 3.12 *Digitaria decumbens* suffering from Pangola stunt virus

(ii) Reaction to environmental extremes

Local conditions will dictate the relative importance of selection for tolerance of frost, heat waves, and fire or of drought or flooding.

Frost is a regular feature of many sub-tropical and some high altitude tropical environments. True frost-tolerance amongst summer growing plants is rare, but good persistence of plant crowns and buried stem buds after frosting of the tops is widely available. We need to distinguish frost escape, which is often associated with winter dormancy, and plant capacity to grow under cold conditions, which often renders a plant with young winter growth more susceptible to frost damage.

Most improved tropical grasses are fire-tolerant; *Melinis minutiflora* is an exception. On the other hand, most herbaceous tropical legumes are damaged by fire. *Stylosanthes guyanensis* cv. Oxley has a buried crown and recovers quickly after firing; the other released varieties of *S. guyanensis* have very different crown characteristics and are easily killed by fire. Fortunately fire stimulates the germination of hard seeds, and seedling regeneration after firing may restore the sward.

Considerable variation in drought and flood tolerance is available, and these topics will be discussed later.

Shade tolerant grasses are needed for growing under tree crops.

(iii) Tolerance of pests and diseases

The value of animal products from tropical pastures often does not justify the cost of fungicides and insecticides. Also there is a possibility of environmental damage and, in some countries, residues of these compounds in animal products are prohibited. For these reasons pasture scientists concentrate attention on the tolerance of plants to attack, rather than on the chemical control of such outbreaks. Many pest and disease problems are local and infrequent in occurrence. It is best to observe how different cultivars survive attack. The use of compound pasture mixtures including many species provides a broader range of resistance against what are often unknown future problems.

There are obviously very great differences in susceptibility between species. Some entomological problems are associated with plant habit, e.g. funnel ants (*Aphenogaster* sp.) flourish in sod-forming pastures of *Paspalum dilatatum* and *Pennisetum clandestinum*, but have a low density in bunch-type pastures of *Panicum maximum*. Attack of the weevil *Amnemus quadrituberculatus* has been disastrous to *Desmodium intortum* and *D. uncinatum*, where *Macroptilium atropurpureum* and *Glycine wightii* have survived well. *Macroptilium atropurpureum* suffers severe attacks of *Rhizoctonia solani* under humid conditions where *Centrosema pubescens* is untroubled.

Disease resistance is the main consideration in some selection programmes. For example, variation in resistance of *Stylosanthes humilis* to *Colletotrichum* sp. is evident in Florida. The University of Florida has released *Digitaria decumbens* cv. Transvala, which is resistant to the pangola stunt virus which has devastated the common form of *D. decumbens* in some areas of Latin America.

(iv) Compatibility in mixtures

There are no problems of compatibility where monospecific pastures are desired, e.g. for high nitrogen input grass pastures or special purpose legume swards. In these cases the farmer looks for the most vigorous plant in that situation which will successfully dominate and suppress its competitors.

For mixed pastures, it is difficult to lay down the guidelines which result in a successful, stable marriage of species. In some cases there are features of the association which provide compatibility. For example, species with different rooting depth may exploit different layers of the soil profile; the deep-rooted *Medicago sativa* is probably more compatible with *Panicum maximum* var. *trichoglume*, which has many shallow roots, than with the deep-rooted *Cenchrus ciliaris*. In moist environments with seasonal temperature variation, species may exhibit complementary growth rhythms which minimise competition. In grass/legume mixtures, a cyclic pattern of alternating dominance may be associated with fluctuations in nitrogen supply; grass nitrogen starvation leads to legume dominance and improved nitrogen supply, which in turn leads to grass dominance.

Some pasture scientists have attempted to characterise the types of plant habit which lead to compatibility. It is argued that twining legumes such as *Centrosema pubescens* are most compatible with bunch grasses such as *Panicum maximum* which provide both structural support and space between the tussocks for legume growth. At Makerere, Uganda, *Desmodium intortum* was more successful with *Setaria anceps* and with *P. maximum* than with the stoloniferous *Chloris gayana*. Conversely, erect bushy legumes such as *Stylosanthes guyanensis* are better able to dominate low growing creeping grasses like *Cynodon dactylon*. There are so many exceptions to these generalisations that wide validity cannot be claimed. Dominance reflects the superiority of one cultivar over another in a specific environment under specific grazing, cutting and fertiliser management. Co-dominance, or good compatibility, reflects equal growth potential under these conditions, which the farmer can adjust to favour one variety or the other, once he understands the factors which control their growth.

Some illustrations of the genetic variation available from the principal planted tropical pasture species with respect to plant habit, and to climatic and edaphic adaptation are given in the next sections of this chapter.

2 Perennial pasture grasses for the humid tropics

Well adapted pasture grasses are available for all regions of the humid tropics. Varieties can be planted which are readily established, which have high growth rates, which persist well under heavy grazing, and which resist weed invasion. The principal difficulties lie in maintaining these grasses at a high nutritive value, and in obtaining reliable supplies of high quality seed.

(a) Plant habit

Great diversity of plant habit is available, *Pennisetum clandestinum* is a typical short, dense sod-former; it forms both rhizomes and stolons and is very resistant to heavy grazing. Some creeping forms e.g. *Cynodon plectostachyus*, *Digitaria decumbens*, form very long stolons which give quick ground cover from widely spaced planting centres. *Cynodon dactylon* cultivars differ in their sod-formation; cv. Coastal has many rhizomes, cv. Midland has some rhizomes, cv. African is mostly free of rhizomes and cv. Coastcross forms a much more open, tufted sod with less underground material. *Brachiaria mutica* is a coarse, lax plant with decumbent stems which root at the nodes. *Paspalum dilatatum* and *P. notatum* spread by short rootstocks and *P. plicatulum* has a similar spreading habit but more ascendant culms. *Melinis minutiflora* is a straggling grass with a poorly developed crown and elevated stem apices.

Setaria anceps varies from erect, bunch type forms with a discrete, well defined crown, to short forms with short rhizomes. Most of the *Panicum maximum* varieties are

fig. 3.13 Identical twin zebus on *Cynodon dactylon* at Muguga, Kenya

fig. 3.14 *Melinis minutiflora* growing in Laos

erect bunch grasses with a distinct crown. *Pennisetum purpureum* is probably the tallest forage grass in use.

(b) Breeding systems

Some varieties are sterile and form no seed. *Digitaria decumbens* is a sterile triploid; *Cynodon dactylon* cv. Coastcross −1 is a completely sterile hybrid. Many tropical grasses produce seed by a process known as apomixis. In this process, seed is produced without union of the male and female gametes, embryos develop mitotically from mother tissue, and genetic constitution of the seed is true to the mother plant. This results in the maintenance of higher varietal purity. Examples of plants with apomictic breeding systems are *Brachiaria decumbens, Melinis minutiflora, Paspalum plicatulum,* whilst *Panicum maximum, Paspalum dilatatum* and *P. notatum* are mainly apomictic.

Cross-pollinating species are less genetically stable. *Setaria anceps* is cross-pollinating with limited self-fertility. *Pennisetum purpureum* has some apomictic lines but is mainly cross-breeding, and is propagated vegetatively.

(c) Variety development

Plant selection by scientists and farmers has given rise to distinct varieties suited to particular district and farm situations; this has now occurred within most of the major species used. *Panicum maximum* is used as an illustration of the diversity recognised (Table 3.8).

West African forms, such as cv. Hamil and cv. Coloniao (grown widely in Brazil), are very tall, bulky varieties; the presence of thick, tall stem residues poses a management problem. cv. Makueni and the common Queensland variety (cv. Riversdale) are medium height forms. Queensland material has been very successful over a wide range of tropical conditions; cv. Makueni has better cool season growth but superior animal production has yet to be demonstrated. cv. Gatton is a lower growing, leafy form used in the sub-tropics. Gatton and Queensland varieties are better adapted to drier conditions than the other varieties.

All the above forms are bunch grasses; cv. Embu is a stoloniferous type with inferior yield under Queensland conditions.

(d) Climatic adaptation

(i) Daylength

The length of day (or more exactly, the length of night) decides whether flowering occurs in many grasses. Most tropical plants flower in response to shortening days, e.g. *Paspalum plicatulum* cv. Rodds Bay grown at Brisbane (lat. 28°S) initiates hidden flowers within the leaf sheath each year about February 14 when daylength reaches 13·1 h. Many plants have 'critical' daylengths (below which flowering occurs) with high enough values for flowering to proceed in the tropics whenever growing conditions are favourable.

A few plants, e.g. *Paspalum dilatatum* and *P. notatum* have been listed as long-day plants; they flower in the early summer and flowering is hastened by warm temperatures.

(ii) Temperature

Tropical grasses are all adapted to warm environments. Maximum growth rate usually occurs at high temperatures, about 35°C, which is about 5°C higher than the optimum for most of the tropical pasture legumes. These figures might be contrasted with the optimum temperature range of about 15 to 20°C for many temperate grasses and legumes. Digestibility decreases as temperature increases so the very high growth rates of tropical grasses under hot conditions are not such a great advantage as may be thought.

Pasture varieties are needed for environments which range from high altitude or subtropical districts with cold winters to lowland equatorial regions with consistently

Table 3.8 Characteristics of *Panicum maximum* varieties.

Variety	Origin	Maximum Height (m)	Description
Hamil	W. Africa?	4	Blue green leaves; broad leaf blades softly hairy, but less hairy than Queensland variety; stiff hairs on the basal leaf sheaths
Coloniao	W. Africa?	3·5	Blue green leaves; broad, glabrous leaf blades; soft short hairs on the nodes; thick, fleshy stems
Makueni	Kenya	2	Leaf blades, leaf sheaths, stems and nodes covered with short dense hairs; spikelets purplish and covered with short dense hairs
Queensland	E. Africa	2	Green leaves; leaf blades and leaf sheaths narrower and thinner than Hamil or Coloniao; hairy nodes
Gatton	S. Rhodesia	1·7	Green leaves; finely pubescent leaf sheaths, glabrous nodes

NB 1 ft = 0·254 m.

warm conditions. For example, creeping grasses adapted to light soils in southern USA range from *Cynodon dactylon* cv. Midland, to cv. Coastal, to cv. Coastcross – 1 to *Digitaria decumbens* as one moves south from Tennessee to south Florida. *Paspalum dilatatum, P. notatum, Pennisetum clandestinum, Setaria anceps* are grasses possessing some cold tolerance, whilst the distribution characteristics of the tall *Panicum maximum* varieties are more truly tropical. *Digitaria decumbens* ceases growth when night temperatures fall to 11°C, irrespective of the level of day temperatures. This is due to the continued accumulation of starch in the leaves during cool nights, which causes injury to the chloroplasts on which photosynthesis depends.

(iii) Moisture

The grasses discussed in this section are adapted to the humid tropics. These are variously defined, and may be broadly considered as districts receiving more than 1 000 mm annual rainfall with at least seven or eight months effective rainfall for pasture growth. The effectiveness of annual rainfall depends upon the rate of losses from evaporation, run-off or percolation and the water storage capacity and availability characteristics of the soil. It is also influenced by the rainfall distribution pattern with respect to frequency and to temperature optima for growth.

Examples of grass adapted to the lower rainfall regions of the humid tropics, or where a pronounced wet/dry season climate prevails, or to dry soil sites, are *Paspalum plicatulum, Setaria anceps* cv. Kazungula, *Digitaria decumbens*, *Panicum maximum* cv. Gatton, and *Cynodon dactylon* cv. Suwannee.

(e) Edaphic adaptation

As indicated earlier, many of the grasses in commercial use have been selected for their capacity to respond to improved nutrition. The distribution of *Pennisetum clandestinum* is mainly restricted to fertile red latosols and to areas receiving additional nitrogen inputs, e.g. near cattle yards and in night grazing fields. Table 3.9 illustrates how sensitive to nitrogen supply is the competitive capacity of *P. clandestinum*. These data were collected from a tableland area near Atherton, Queensland (lat. 17°S), after three years of fertiliser application. Three per cent of the control sward was *P. clandestinium* which attained co-dominance with *P. dilatatum* with inputs of 224 kg nitrogen per ha per year and represented 76 per cent of the sward at the highest nitrogen rate used. Conversely, the content of *Axonopus affinis* was lowest at high rates of nitrogen supply, varying from 87 to 1 per cent of the botanical composition of the sward.

Under high rainfall (3 200 mm) coastal lowland con-

fig. 3.15 *Brachiaria ruziziensis* growing in Bali

ditions at South Johnstone (lat. 18°S), *Brachiaria decumbens* has been the most productive grass with high inputs of nitrogen. Under sub-tropical conditions, *Setaria anceps* is highly responsive to nitrogen, and has a higher potassium requirement than other grasses tested. *Hyparrhenia rufa* is well adapted to low fertility situations in Central and South America, and in Central Africa. *Melinis minutiflora* has a similar reputation. On one very low fertility soil in central Laos, *M. minutiflora* swards yielded 11 700 kg per ha with 3 per cent volunteer weeds, where other grasses such as *Panicum maximum* were lower yielding and unable to exclude weed invasion.

This topic is explored more fully in Part 5.

Plants may also have to be chosen for areas where there is a high water table during part or all of the year.

Table 3.9 Nitrogen supply and botanical composition of mixed swards. (*Source:* Gartner, J. A. (1966). *Proc. 10th Internat. Grassld. Congr.* Helsinki p. 123.)

Rate of Nitrogen Application (kg/ha)	***Pennisetum clandestinum*** (per cent)	***Paspalum dilatatum*** (per cent)	***Axonopus affinis*** (per cent)
0	3	10	87
55	10	21	69
112	12	33	55
224	44	46	11
448	76	23	1

NB 1 lb per ac = 1·12 kg per ha

Sometimes plants may be required for soils that are actually flooded. *Brachiaria mutica, Leersia hexandra* and *Entolasia imbricata* perform well in swamp situations. *Echinochloa polystachya* from Colombia gives good summer production on many Latin American sites, and tolerates wet, non-saline conditions. *Echinochloa pyramidalis* is successful under flooded conditions in the Amazon basin. In south Florida *Hemarthria altissima*, collected from South Africa, is showing promise in both wet and saline conditions. *Acroceras macrum* will also tolerate a high water table, but its growing season is limited. *Digitaria decumbens* is often planted in poorly drained sites. On the other hand *Melinis minutiflora* and *B. decumbens* require well drained conditions.

The establishment of these plants is considered in Part 4, and grazing management is discussed in Part 5.

3 Perennial pasture grasses for the dry tropics

Sowings of improved pasture grasses are more hazardous in the sub-humid and semi-arid regions of the tropical world; establishment of the sward is less reliable and responses to applied fertiliser are less likely to be profitable. However, selected species of higher nutritive value than unselected native species are available, and these species when sown are tolerant of heavy grazing and persistent un-

fig. 3.16 *Brachiaria decumbens* growing under high nitrogen conditions in North Queensland

fig. 3.17 *Brachiaria mutica* growing in a wet situation below the spillway of a dam

der drought conditions. Productivity gains are therefore possible if the native species are replaced with well adapted sown species. It will often be advisable to concentrate the grass improvement programme first on the higher fertility soils.

(a) Plant habit

Urochloa mosambicensis is an example of a low-growing mat-forming species which has shown considerable promise in the dry tropics of southern Africa and northern Australia. *Cenchrus ciliaris* forms vary from short, bunch grasses to strongly rhizomatous types with erect culms. The short-lived *Sorghum almum* is a very tall, erect free-seeding plant which forms strong rhizomes. *Chloris gayana* cultivars vary in their stoloniferous development. *Panicum coloratum* var. *makarikariense* has decumbent stems which root at the nodes.

Panicum maximum var. *trichoglume* cv. Petrie and *P. antidotale* are examples of erect, tussock forming grasses.

There are very real differences in the type and extent of

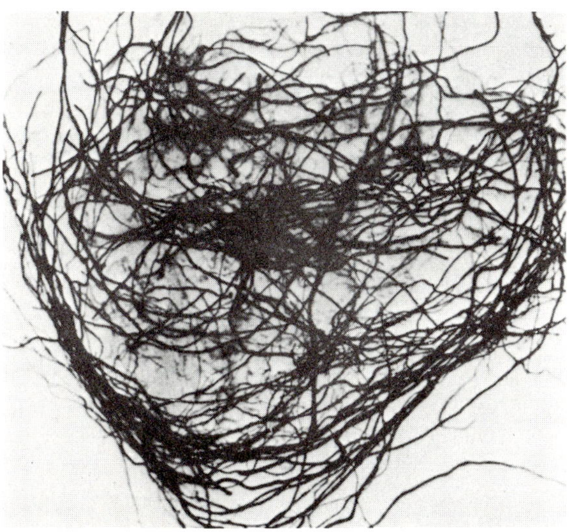

fig. 3.18 Coarse, massive roots of *Cenchrus ciliaris*

fig. 3.19 Fine, richly branched roots of *Panicum maximum* var. *trichoglume*

Table 3.10 Varietal characteristics of *Cenchrus ciliaris*. (*Source:* Humphreys, L. R. (1967). *Trop. Grass.* 1:123.)

Variety	Height (mm)	Stem Diameter (mm)	Main Shoot Density (no./m^2)	Percentage Shoots Vegetative	Percentage Leaf
West Australia	450	0·92	1 620	35	40
Cloncurry	600	1·20	850	50	45
Gayndah	600	1·31	1 030	94	45
Boorara	850	1·50	350	85	52
Biloela	990	1·78	380	65	47
Q.1004 (similar to cv. Molopo)	1 050	1·45	300	99	54
Tarewinnabar	1 250	1·74	450	87	40

the root systems of these plants. For instance, *Cenchrus ciliaris* has a much deeper root system with coarser, less branched roots than *P. maximum* or *C. gayana.*

(b) Breeding systems

Apomixis is the predominant form of seed production, as for the grasses of the humid tropics. The discovery of a fertile female plant in *Cenchrus ciliaris* has made hybridisation programmes possible with this species, since male pollen from the apomictic lines can be used to fertilise this rare fertile female line, with consequent release of much variation. *C. gayana* and *S. almum* are cross-fertilised. *P. coloratum* is mainly cross fertilised and self-incompatible; planting material for vegetative propagation should be taken from several clones, or seed may not be produced.

(c) Variety development

C. ciliaris is taken as an example of a species from which many selected varieties have been released. Table 3.10 summarises some measurements made in February at the Brian Pastures Research Station, Gayndah, Queensland (lat. 26°S). At this time the purple-seeded short West Australia variety was 450 mm high, the mid-height Cloncurry and Gayndah varieties 600 mm, and the tall varieties about 850 to 1 250 mm. The short varieties are characterised by a high shoot density and slender stems; the tall varieties are rhizomatous with low shoot density and coarse stems. All varieties are relatively early flowering in this environment, but plants with a high degree of flowering (i.e. per cent shoots not vegetative) usually have less leaves. The tall varieties generally yield more dry matter and compete more successfully with invading native species than the short varieties. The high seed yields of the short varieties probably contribute to their success in semi-arid regions; seedling drought survival is superior in the taller varieties Biloela and Gayndah.

(d) Climatic adaptation

(i) Temperature

P. maximum cv. Petrie, *C. gayana*, and *P. coloratum* are usually believed to make more growth under cool conditions than *C. ciliaris.* Frost kills the top growths of these plants but they can still survive a medium frost. *P. coloratum* forms more new basal tillers during the cool season than the other species mentioned.

(ii) Drought tolerance

Considerable evidence has accumulated to indicate that most varieties of *C. gayana* which are also grown in the humid tropics, are less drought tolerant than *P. maximum* cv. Petrie, whilst *C. ciliaris* will succeed in drier situations than either of these species. In mixed swards the dominance of one of these species is often the result of recent rainfall conditions, *C. gayana* predominating in wet years and the other species in dry years.

(e) Edaphic adaptation

In East Africa, a CSIRO. plant collecting expedition

fig. 3.20 *Cenchrus ciliaris* making vigorous growth after the breaking of a long drought at Gayndah, Queensland

noted that no natural spread of *Cenchrus ciliaris* occurred on soils which had a surface pH below 7·0. Most plants were found in areas having surface soil pH of 7·0 to 7·5. Queensland Department of Primary Industries work has shown some differences in variety tolerance of low pH, cv. American being superior in this respect.

Some of the species selected come from the African floodplains, whose fertility is constantly renewed by flooding. *C. ciliaris* has a high phosphorus content, and has shown superior competitive ability on soils of better phosphorus status. In Part 4, an illustration is given in Table 4.9 of how the capacity of *C. ciliaris* to compete with the native *Triodia pungens* is improved by fertiliser addition.

P. maximum cv. Petrie is very sensitive to nitrogen deficiency and competes best under high fertility conditions. In one experiment at the Brian Pastures Research Station,Gayndah, Queensland (lat. 26° S) varying levels of nitrogen fertiliser, from 0 to 224 kg nitrogen per ha, were applied to an old sward of *P. maximum* cv. Petrie which had been invaded by *Bothriochloa bladhii* and *Dichanthium sericeum.* The content of *P. maximum* cv. Petrie after one season of fertilising and lenient defoliation varied from 50 per cent in the no-nitrogen plots to 100 per cent in the plots receiving 130 kg nitrogen per ha or more.

Chloris gayana, P. coloratum and *S. almum* have good tolerance to salinity, a problems which often develops in cleared inland areas. *Cenchrus ciliaris* shows only moderate salinity tolerance. Under conditions where salinity levels in the soil fluctuate seasonally, the more productive *C. ciliaris* varieties, such as Biloela and Gayndah, are to be preferred. Fig. 3.21 shows how the yield of grass is negatively related to salt concentration. The concentration of salt (milli-equiv. per litre) which reduced growth potential of a particular variety by one-half was as follows: American 28, Tarewinnabar 44, Gayndah 66, Biloela > 110, and Molopo > 125. However, the varieties Biloela and Tarewinnabar showed a higher growth rate under most conditions.

C. ciliaris is intolerant of poor drainage situations. Some sowing sites in valley bottoms are subject to intermittent short duration flooding. *P. antidotale, P. maximum* cv. Petrie, and *Urochloa mosambicensis* are particularly susceptible to this hazard. *Chloris gayana* and *Cenchrus ciliaris* show moderate tolerance; cv. Tarewinnabar is the best *C. ciliaris* variety tested for this situation. *P. coloratum* var. *makarikariense* shows good tolerance of flooding.

Flood damage is lessened if plants are dormant. Tall plants or leniently grazed plants are less likely to be fully submerged. Flowing water leads to better aeration than still water. Flood damage is worse under hot conditions, when scalding may occur and when the plants' oxygen demand is greater. Different varieties have varying flood tolerance which is probably related to plant capacity to accumulate malic acid rather than ethanol under anaerobic conditions.

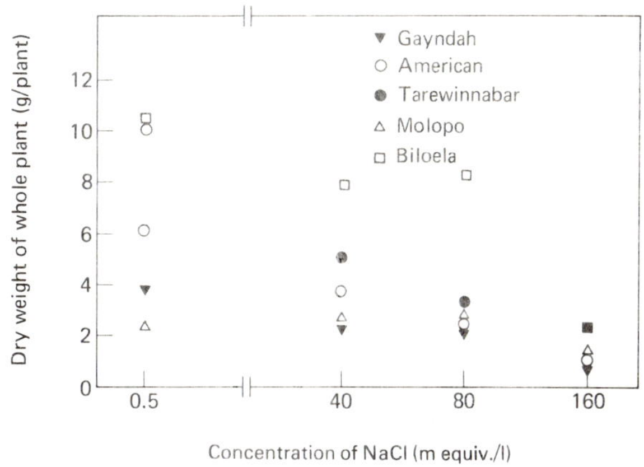

fig. 3.21 Growth of *Cenchrus ciliaris* cultivars and salinity level. (*Source:* Graham, T. G. and Humphreys, L. R. (1970). *Aust. J. Exp. Agric. Anim. Husb.* **10**:725).

4 Perennial pasture legumes for the tropics

Perennial pasture legumes give a longer sequence of feed production than annual legumes, and are also usually capable of fixing larger amounts of nitrogen than annual legumes. Their use is indicated for long term pastures where rainfall is sufficient to permit persistence. The varieties in commerce have been drawn from many countries, mainly in South America and Central Africa, and hence are available for quite diverse habitats. Improved persistence under heavy grazing and improved energy value are the main objectives.

(a) Plant habit

A restricted number of tropical legumes are truly procumbent and are as tolerant of heavy grazing as creeping grasses such as *Digitaria decumbens. Desmodium heterophyllum* and *D. ovalifolium* fall in this class, but unfortunately their poor seed production characteristics only fit them for vegetative propagation. *Trifolium repens* and *T. semi-pilosum* are highly stoloniferous, and are suited to high altitude tropics and to the moist sub-tropics. *Lotononis bainesii* is also stoloniferous, and underground crown development (see Fig. 3.22) helps its persistence under grazing. *Stylosanthes guyanensis* cv. Oxley and the promising species *S. hamata* also develop an underground crown. These plants escape shading if grazing pressure is maintained.

fig. 3.23 Procumbent *Trifolium semi-pilosum*, with broad white mid-rib

Many tropical pasture legumes are trailing and twining with stems which root at the node when the stems are close to the ground; however they have elevated growing points when climbing up companion grasses. They escape shading if lightly grazed, and have a good capacity to smother weeds. *Centrosema pubescens, Macrotyloma axillare, Glycine wightii, and Pueraria phaseoloides* are examples; *P. phaseoloides* has a more superficial root system. *Macroptilium atropurpureum* cv. Siratro was bred from two Mexican ecotypes. It has a much more stoloniferous character; in one test the parents had a stolon density of 6 and 30 per m^2 whilst the released cultivar had 75 stolons per m^2. *Desmodium intortum* and *D. uncincatum* are semi-decumbent plants which do not twine, but which are carried upwards by companion grasses; their

fig. 3.22 Well grazed *Lotononis bainesii*, showing creeping stolons and strong tap-root

fig. 3.24 *Desmodium heterophyllum* creeping amongst heavily grazed *Digitaria decumbens.* Coin is 3 cm diameter

fig. 3.25 *Pueraria phaseoloides* smothering all weeds at Borabue, north-eastern Thailand

stems will root if in contact with moist soil. *D. heterocarpon* shows great promise in south Florida and may be more persistent under grazing than *D. intortum* and *D. uncinatum.*

Erect legume species which grow tall also have a competitive advantage when growing with a companion grass; *Stylosanthes guyanensis* cv. Schofield is an example. Varieties of *Medicago sativa* currently grown in the subtropics and high altitude tropics have erect stems and very strong tap-root development.

The use of planted shrub legume pastures has increased in recent years, e.g. *Sesbania grandiflora* is widely grown in Southeast Asia as a legume tree from which fodder is cut. Shrubs have several advantages over the smaller herbage legumes. Shrub legumes such as *Leucaena leucocephala* are more deeply rooted than many herbage legumes, exploiting deeper reserves of soil moisture, and maintaining green foliage throughout the dry season. A common plan in the wet/dry season tropics is to feed cattle grass at the beginning of the dry season, and to switch to shrub foliage later in the dry season. Finally, in abnormally long, dry seasons tree foliage is fed. In the Sahel and Sudanese zones of Africa it is normal practice for herdsmen to cut fodder for their livestock from legume trees each dry season. In cold regions, green leaf will be borne above the height at which light radiation frosts occur. Shrub legume dominance is more easily achieved in mixed pastures, and once the plants are well established, some of the canopy cannot be grazed and this helps ensure plant persistence. Some scientists believe the deep light profile of shrub/grass pastures provides more efficient conversion of radiant energy. Finally, shrubs can be used as a 'living fence'.

fig. 3.26 *Macroptilium atropurpureum* cv. Siratro. Note the lobed leaves

The principal difficulty in using shrub pastures is to maintain a level of use which prevents a forest developing, with forage quite inaccessible to stock. However, in this eventuality the forage can always be mechanically lowered to the animals with axe, chain saw or bulldozer, and treated as a feed reserve for times of scarcity.

Some herdsmen regard the inaccessibility of forage during the growing season as an advantage since it ensures that reserve feed is available for cutting in the dry season.

(b) Breeding systems

Almost all the commercial species, e.g. *Centrosema pubescens, Glycine wightii, Leucaena leucocephala, Lotononis bainesii, Macroptilium atropurpureum, Macrotyloma axillare, Pueraria phaseoloides* and *Stylosanthes guyanensis*, are self-fertilised, and exhibit good varietal stability in consequence. Some are also cleistogamous, i.e. fertilisation takes place within the closed flower bud. *Trifolium repens* and *Medicago sativa* are cross-pollinating with a high degree of self-compatibility; tripping of the flowers by bees improves seed-set.

(c) Variety development

Variety development has taken place in most species; one exception is *Lotononis bainesii*, in which little variation is evident. *Stylosanthes guyanensis* is an illustration of a species in which cultivars have been established and in which the recognition of considerable variation (especially

fig. 3.27 *Desmodium intortum* flowering

fig. 3.28 Segmented racemes of *D. uncinatum*

with respect to grazing tolerance and yield persistence) is likely to lead to new cultivars being released.

S. guyanensis cv. Schofield was the first variety commercially grown in Australia. It is an erect, vigorous, thick stemmed form from Brazil. It behaves well in pioneer situations, being easy to establish and smothering weeds well if grazing is lenient. Good nodulation usually occurs from local rhizobial strains. Schofield flowers very late in the growing season, in response to shortening days.

cv. Endeavour is a more recent release, collected from south-west Guatemala (lat. 14°N) at 1 850 m in a region with 2 100 mm annual rainfall and a definite dry season. This variety is semi-erect, more branched, with thinner, more pointed, narrower lighter green leaves than cv. Schofield. It has excellent establishment characteristics and has shown better early summer yields than Schofield at a number of sites. It flowers about two weeks earlier than cv. Schofield at lat. 18°

cv. Cook was introduced from a site at 1 400 m in Colombia (lat. 4°), where the annual rainfall is about 4 000 mm; however, it grows well at drier sites. It is a rather erect variety, with narrower, more pointed leaves than Schofield; they are blue-green with bright red stipules clasping the stem. It flowers about six weeks earlier than cv. Schofield at lat. 18°, and the flowers have an orange coloured standard, rather than pale-yellow, as in cv. Schofield and cv. Endeavour. Cook has better cool season growth and better resistance to *Botrytis* than other stylo varieties.

cv. Oxley is a very distinct form from Paraguay with thin stems, small leaves, and a buried crown. It flowers in response to long days, in contrast to the other three varieties mentioned, and flowers are produced over much of the growing season. It differs from the other three varieties in having a specific rhizobial requirement for inoculation. cv. Oxley is adapted to drier and frostier situations than the other varieties.

(d) Climatic adaptation

(i) Daylength

Most of the perennial tropical legumes respond to short-days. However, considerable variation is available in the requirements for flowering. Fig. 3.29 illustrates this for 120 introductions of *S. guyanensis* grown at Lansdown, (lat. 20°S). First flowering time for different lines occurred when day-length varied from 13·2 hours to 10·8 hours; however the great majority of plants flowered in response to days shortening to less than 12·0 hours. *S. guyanensis* is unusual amongst tropical legumes in possessing long-day forms, e.g. cv. Oxley, which will not flower in short days. Some long-day plants from temperate regions, e.g. *T. repens* and *Medicago sativa* are grown in the tropics, but obligate long-day forage varieties originating in the tropics are few.

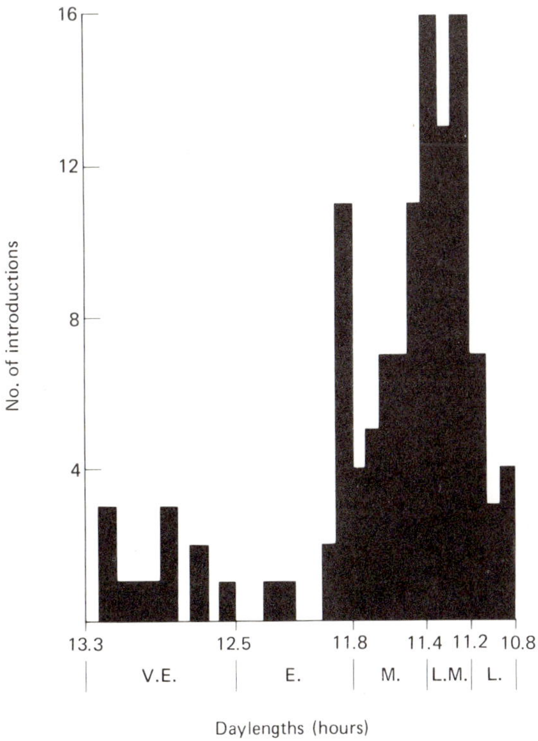

fig. 3.29 Distribution of mean daylength at flowering at 'Lansdown', (lat. 20°S) for 120 introductions of *Stylosanthes guyanensis*. Maturity types represented as follows: V.E. very early; E. early; M. mid-season; L.M. late mid-season; L. late. (*Source:* Edge, L. A. et al (1973). *Aust. J. Agric. Res.* **24**:511).

fig. 3.30 Buffalo and native cattle grazing on erect *Stylosanthes guyanensis* dominant pasture at Borabue north-eastern Thailand

Glycine wightii, Centrosema pubescens, Desmodium intortum and *D. uncinatum* are plants with a clear-cut, obligate short day flowering response. On the other hand, some forage plants are in inductive flowering conditions for most of the year in the tropics and sub-tropics. Flowering may be hastened by short days, but the plants will produce flowers whenever shoots are old enough and large enough. *Macroptilium atropurpureum* cv. Siratro behaves in this fashion.

(ii) Radiation

Legumes are generally shade-intolerant. Differences in shade-tolerance between tropical pasture legumes are not well documented, although there is current research in this area. This question has particular importance for grazing management, since grasses tend to grow faster than legumes and shade them, and for the use of pastures under

fig. 3.31 Flowering *Centrosema pubescens* with *Paspalum plicatulum*

fig. 3.32 *Macrotyloma axillare* afflicted with little-leaf virus

companion tree crops. One of the most promising fields for pasture development is in plantations of coconut, oil palm, kapok, and other crops. Some of the perennial pasture legumes originated as plantation cover crops, intended to control weeds and erosion; grazing capacity was not a consideration. This is perhaps well reflected in *Calopogonium mucunoides*, a good cover crop poorly accepted by animals. *Pueraria phaseoloides* and *Centrosema pubscens* have a reputation for better shade tolerance than other tropical pasture legumes.

(iii) Temperature

As indicated in a previous section, most tropical pasture legumes have a temperature optimum for growth of about 30°C. *S. guyanensis* cv. Schofield has a slightly higher temperature optimum. *Centrosema pubescens* and *Pueraria phaseoloides* also have the reputation of being hot climate species, making very little growth under cool conditions. In southern Queensland, *C. pubescens* will only persist on the warmer, northern facing hillsides and not on the southern slopes. Even the plants which are used extensively in the sub-tropics, e.g. *D. intortum D. uncinatum*, and *Macroptilium atropurpureum*, make almost no growth at a temperature regime of 18° day/13°C night.

One experiment showed that *Glycine wightii* cv. Cooper had a temperature optimum of 27°/22°C. The late flowering *G. wightii* cv. Tinaroo makes good cool season growth on the Atherton Tableland, Queensland, 45 kg per ha per day being recorded. In this plant, whose terminal shoots remain vegetative, leaf differentiation continues on existing branches during early flowering. *Trifolium repens* and *Lotus major* make best growth at about 24°C, and these species do not give their optimum performance under the higher summer temperatures of the continental tropical climates.

None of the pasture legumes of tropical origin have any true frost tolerance, with the exception of *Lotononis bainesii*.

(iv) Moisture supply

Most of the commercially available perennial pasture legumes require an annual rainfall of about 1 000 mm and a pasture growing season of at least seven to eight months if good levels of growth and nitrogen fixation are to occur. Successful persistence of the more drought tolerant cultivars (e.g. *G. wightii, M. atropurpureum, Leucaena leucocephala, Macrotyloma axillare, S. guyanensis* cv. Oxley, *Desmodium sandwicense* and *Clitoria ternatea*) occurs in areas receiving about 750 mm annual rainfall. Some

fig. 3.33 Flowering *Stylosanthes guyanensis* with *Panicum maximum*

Table 3.11 Copper requirements of pasture legumes. (*Source:* Andrew, C. S. and Thorne, P. M. (1962). *Aust. J. Agric. Res.* **13**:821.)

Species	Relative Yield of Control Treatment*
Desmodium uncinatum	60
Macroptilium lathyroides	46
Centrosema pubescens	45
Stylosanthes bojeri	7
Medicago sativa	17

* Percentage yield of plants receiving no copper relative to maximum yield with good copper supply

current breeding programmes, e.g. with *M. atropurpureum*, are directed to improving persistence in the dry tropics, for which few perennial legumes are well adapted. *S. hamata* cv. Verano shows considerable promise for the dry tropics; it has very high seed yields for a perennial species and behaves as an annual in adverse circumstances.

(e) Edaphic adaptation

Many of the tropical legumes are adapted to a wider range of soils and are less demanding in their mineral needs than the temperate legumes. Some species exhibit good tolerance of low pH, low calcium supply, and of excess aluminium and manganese supply. They also show good capacity to extract nutrients from soils of low mineral content. This question, and the related topic of mineral requirements for rhizobial symbiosis, are treated more fully in Part 5.

An example of differences in the capacity of plants to grow under conditions of low nutrient supply is given in Table 3.11. A number of tropical and temperate pasture legumes were grown at varying levels of copper supply. The ratio was calculated of the yield of the plants receiving no additional copper to the maximum yield of plants of the same species receiving good copper supply. *D. uncinatum* showed a superior capacity to extract copper under low supply conditions, the control *D. uncinatum* plants yielding 60 per cent of the maximum *D. uncinatum* yield. On the other hand, *S. bojeri* and *M. sativa* yielded only 7 and 17 per cent respectively of the maximum yields for those species. In the control treatments, copper content of plant tops varied from 1·7 ppm in *M. sativa* to 4·5 ppm in *D. uncinatum*.

Generally 4 to 5 ppm copper content is regarded as marginal for plant copper adequacy. The copper concentrations which gave maximum growth were similar for all species, but some species grew more at lower copper availability. The sensitive species were relatively poor extractors of copper. *S. bojeri* had a reasonable amount of copper uptake per unit weight of root tissue, but its high shoot/root ratio placed it at a disadvantage.

Usually *Stylosanthes* spp. are regarded as legumes quite tolerant of low levels of nutrient supply, and tolerant of acidity. *Lotononis bainesii* is also able to grow on low

Table 3.12 Soil type effects on seedling growth and nodulation of *Glycine wightii*. (*Source:* Thomas, D. and Whiteman, P.C. (1971). *Aust. J. Exp. Agric. Anim. Husb.* **11**:513)

Soil Type	Clay plus Silt (per cent)	Plant Weight (g)	Nodule Weight (g)	Nitrogen Fixation Efficiency*
Black earth	74	15·2	1·6	0·030
Alluvium	35	13·8	1·4	0·034
Krasnozem	81	10·3	0·7	0·039
Yellow podzolic	38	6·6	0·6	0·015
Siliceous sand	14	7·8	0·7	0·037

* Regression coefficient of nodule weight on log plant nitrogen yield

calcium soils. Two tropical species which have a higher requirement for calcium and which are best adapted to less acid soils are *Leucaena leucocephala* and *Glycine wightii*. In East Africa, a CSIRO plant collecting expedition found no naturally occurring *G. wightii* on soils with surface pH less than 6·0, and 69 per cent of collections were made from soils with surface pH in the range of 7·0 to 8·5. By contrast, 82 per cent of *Setaria anceps* collections were made from areas having surface soil pH in the range of 5·0 to 6·5.

Table 3.12 shows the relative growth of *G. wightii* cv. Tinaroo in a University of Queensland pot experiment using five soil types. All known nutrient deficiencies, with the exception of nitrogen, were corrected. It will be noted that poor growth, 6·6 and 7·8 g, occurred on the yellow podzolic and siliceous sands respectively, whilst 15·2 g yield occurred on the black earth. Nodule weight was similarly affected. The efficiency of nitrogen fixation was assessed by comparing the relative plant nitrogen yield with nodule weight; this indicated a low value (0·015 units) for the yellow podzolic soil relative to values of 0·030 to 0·039 units for the other soils. *G. wightii* has a reputation for poor adaptation to podzolics and solodics, but usually grows well on black earths, krasnozems, and well drained alluvials.

None of the commerically available tropical legumes are really well adapted to badly drained situations. Table 3.13 shows the comparative frequency of occurrence of species

Table 3.13 Botanical composition and site effects. (*Source:* Bryan, W. W. (1970). *Proc. 11th Internat. Grassld. Congr.* p. 636)

Species	Percentage frequency	
	Dry Soils	Wet Soils
Desmodium intortum	42	33
D. uncinatum	7	5
Lotononis bainesii	15	4
Trifolium repens	34	50
Chloris gayana	43	17
Digitaria decumbens	50	48
Paspalum commersonii	5	16
P. dilatatum	45	55
Juncus sp.	7	27
Kyllinga sp.	28	49

on wet and dry soil sites at Beerwah (lat. 27°S, 1 630 mm annual rainfall) in the coastal lowlands of Queensland. A wide mixture of species was planted on a very gently undulating site, in which the low-lying wet areas were low humid gley soils and the drier, higher sites were podzolics. The frequencies of *L. bainesii, D. intortum* and *D. uncinatum* were all higher on the dry soils, whilst *T. repens* dominated the wet situations. The grasses *P. dilatum* and *D. decumbens* were well adapted to both wet and dry sites. The perennial legumes which are most intolerant of water logging, and which should be grown on hillsides or well-drained alluvials are: *G. wightii, L. leucocephala, M. atropurpureum, M. axillare*, and *S. guyanensis* cv. Oxley.

5 Self-regenerating annual pasture legumes for the tropics

Self-regenerating annual legumes have some advantages over perennial legumes in climates with a reliable growing season and a clearly defined dry season which is reliably dry. In this type of environment the seed of the annual legume may be a more efficient resting stage to ensure survival than dormant perennial plant crown and root system.

The respiratory energy involved in the maintenance of such a root system is considerable, especially in districts with a long dry season of say, seven months, and much protective tissue against desiccation is also needed to ensure its survival. On the other hand, an early flowering annual will rapidly produce abundant seed during a short growing season, and less of the assimilate produced will be diverted to root tissue, which is inaccessible to the grazing animal. This argument has more force if the seed itself is edible and highly nutritious; for instance, the seed of *Stylosanthes humilis* is well accepted by stock and is licked up from the ground. In a really arid dry season where little deterioration of the dead leaf, stem and inflorescence occurs, animals grazing this pasture will continue to increase in weight.

There are some obvious weaknesses of annual pastures. These plants are unable to respond to rains which are light or out of season since rainfall must be sufficient for the whole cycle of germination and growth to proceed to a point where forage availability is increased. 15 mm of rainfall during the dry season will provide useful green shoots on a perennial, but will cause considerable deterioration in the nutritive value of annual pastures, through leaching and microbiological breakdown. In fact, even a succession of early morning dews will cause rapid deterioration of standing forage. Cold stress may increase the damage done to the forage by this form of breakdown or leaching.

A second weakness of annual pastures is their ecological instability, and the ease with which they are invaded by weeds. A perennial pasture, well managed to maintain its vigour, is better able to use the environmental growth factors to exclude competitors. *S. humilis* pastures are vulnerable to invasion by less acceptable annual grasses, especially after soil nitrogen levels have increased. A third weakness is the vulnerability of annual pastures to erosion damage and excess run-off, especially at the outset of the wet season when ground cover is at its minimum.

These three arguments have much less force if an annual legume is combined with a perennial grass. The perennial can then provide response to out-of-season rain, act as a sink for the nitrogen produced by the legume and exclude weed invasion, and provide rooted ground cover to protect the soil from erosion. However, the maintenance of a perennial grass is easier in an environment with some cool season rains than in a region with a long dry season. In Queensland *Heteropogon contortus* has survived with *S. humilis* under high stocking rates much better at Rodds Bay (lat. 24°S) where there is a cool season rainfall component, than at Lansdown (lat. 19°S) where conditions for the survival of a perennial are less favourable. Discussion in this section will concentrate on *S. humilis,* since it is the most widely used self-regenerating tropical annual legume.

(a) Plant habit

S. humilis usually has procumbent stems and an erect main stem; some variation in erectness of growth exists. Most of the Mediterranean type annual legumes which are used in those parts of the subtropics receiving useful winter rains have a similar habit; *Medicago truncatula, M. scutellata* and *M. denticulata* are examples. *Macrotyloma uniflorum* is a trailing plant which twines on itself and on companion plants; it is especially successful as stand-over cool season forage, leniently grazed during the growing season. *Macroptilium lathyroides* has an erect habit with few basal buds; the tips of the long branches become twining if sufficiently well grown. This plant may behave as annual or biennial.

(b) Breeding systems

S. humilis and *M. lathyroides* are self-pollinating. Most of the *Medicago* species are both self and cross-pollinating.

(c) Variety development

S. humilis exhibits wide variation in plant habit, yield, and seeding characteristics. A variety should be used with flowering characteristics suited to the length of growing season in the district where it is sown. A very late-flowering variety grown at a site with a short growing season may fail, since inadequate seed will be produced to ensure natural regeneration at the start of the next growing season. Conversely, an early flowering variety grown in a long season district may not fully exploit the growth potential of this site. An annual plant, by definition, dies when it has completed its flowering cycle. This problem can be overcome if farmers use seed grown locally or grown in districts with a similar growing season. Genetic shift will also assist the process of local adaptation.

Three varieties were released in Queensland. These were selected by CSIRO for their high yielding characteristics, erect habit, and range of flowering behaviour. They were collected from areas in Australia where they had become naturalised, after prior accidental introduction from South America. cv. Paterson is the earliest flowering, with a critical photoperiod of about 13 hours (i.e. in most environments the plants will flower in daylengths shorter than 13 hours, if the plants are old enough and large enough). This variety was collected at Greenvale, near Charters Towers (lat. 20°S). It is suited to districts having about 550 to 900 mm annual rainfall. The single seed inside the hooked pod has an unusual purplish-black seed coat, in contrast to the brown seed coats of the other varieties.

cv. Lawson is mid-season flowering, and is suited to districts with annual rainfall of about 900 to 1 100 mm. It was

fig. 3.34 Showing the branching habit of *S. humilis*

fig. 3.35 Each white peg marks an introduced plant of *S. humilis* at Fort Pierce, Florida. Note the variation in plant size

collected at Galbraith Station (lat. 16°S) on Cape Yorke Peninsula, Queensland. It forms a particularly high percentage of hard-coated seed, a good survival mechanism, and much of this softens during the dry season.

cv. Gordon is the latest flowering variety, with a critical photoperiod in most environments between 12 and 11·5 hours. It flowers four to six weeks later than cv. Lawson at lat. 19°. It is best used in districts with a wet/dry climate and annual rainfall in excess of 1 100 mm. It was collected from Darwin, Northern Territory (lat. 12°S). cv. Kulumbaru is another late seeding cultivar.

(d) Climatic adaptation

(i) Daylength

As indicated earlier, *S. humilis* flowers in response to short days. Table 3.14 illustrates this response for different selections, grown in controlled environment chambers where the temperature was either 18°C or 30°C constant, and

Table 3.14 Daylength and temperature effects on flowering time of *Stylosanthes humilis* selections. (*Source:* Cameron, D. F. (1967). *Aust. J. Exp. Agric. Anim. Husb.* **7**:489.)

Temperature (°C)	Daylength (hours)	Days to First Flower			
		Selection Number			
		25	78	22	10
18	8	51	50	53	51
	10	50	50	51	58
	12	56	66	58	**
	13	50	74*	59	—
	14	**	74*	—	—
	16	—	**	—	—
30	8	24	26	24	28
	10	23	21	21	22
	12	44	29	41	—
	13	55*	—	—	—
	14	—	—	—	—
	16	—	—	—	—

* Some plants not flowering

** Flowers initiated in some or all plants at end of experiment daylength varied from 8 to 16 hours. It will be seen that short days usually hastened flowering, e.g. selection no. 25 grown at 30°C took 24 days to flower in 8-hour days, but 44 days to flower in 12-hour days. The length of the critical photoperiod varied between selections. For example, at 30°C selections no. 25, 78 and 22 all flowered in 12-hour days, but selection 10 flowered only in 10-hour days or less. It will also be noted that for some selections the critical photoperiod was longer under cool conditions. For example, some line 25 plants flowered in 14-hour days at 18°C, but none flowered in 14-hour days at 30°C. This agrees with a common observation that some plants which flower only in less than 12-hour days in lowland tropics and subtropics will flower in 12-hour days in cooler highland equatorial regions.

The flowering of *S. humilis* under field conditions has been shown to be controlled primarily by day-length, and latitudinal effects account for much of the field variation in flowering time observed in natural populations of *S. humilis*.

(ii) **Radiation**

S. humilis, in common with other annual legumes, is shade-intolerant. This is illustrated from an experiment summarised in Table 3.15. *S. humilis* yielded 13·9 g in full sunlight, but reduction to only 74 per cent sunlight decreased plant yield to 7·4 g. Further shading to 38 per cent sunlight reduced plant yield to 3·0 g and caused 33 per cent mortality. Similarly, further Queensland Department of Primary Industries work showed that where *Heteropogon contortus* and *S. humilis* were growing in the absence of root competition, the legume yield was affected by frequency of cutting the grass, which varied the grass shading effect. Grass uncut for 14 weeks reduced the associated *S. humilis* yield by 40 per cent, compared to more frequent cutting.

Table 3.15 Shading and growth of *Stylosanthes humilis*. (*Source:* Sillar, D. I. (1967). *Qld. J. Agric. Anim. Sci.* **24**:237.)

Shading Treatment (per cent full daylight)	**Yield** (g)
100	13·92
74	7·40
54	6·13
38	3·00

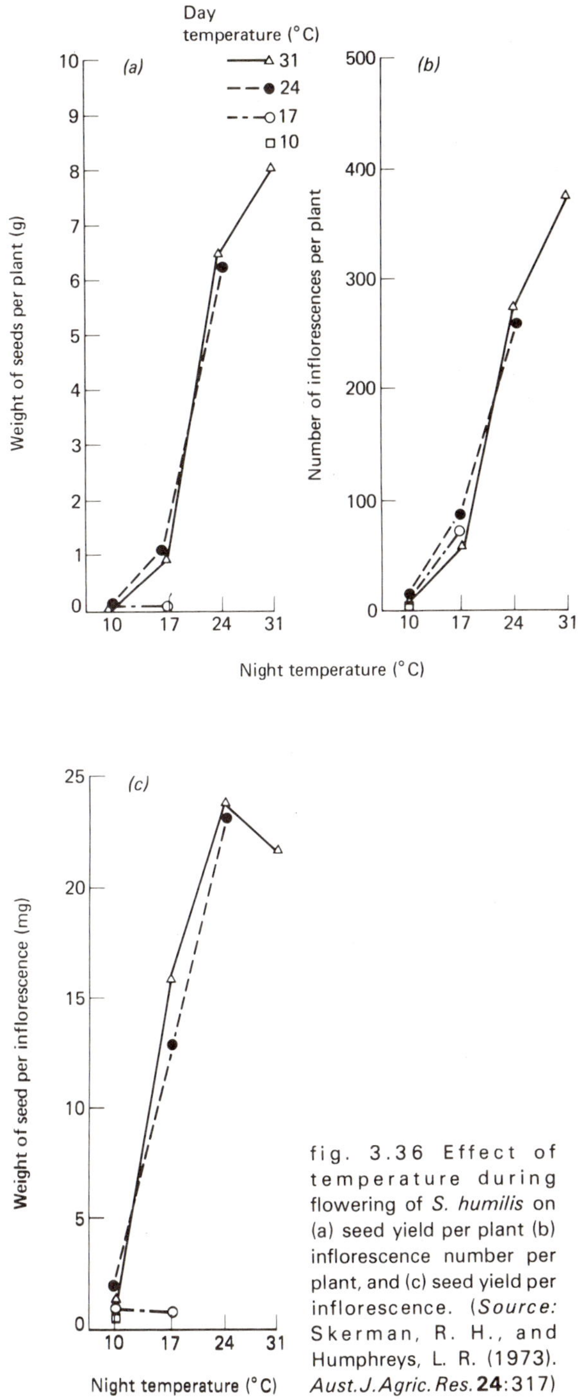

fig. 3.36 Effect of temperature during flowering of *S. humilis* on (a) seed yield per plant (b) inflorescence number per plant, and (c) seed yield per inflorescence. (*Source:* Skerman, R. H., and Humphreys, L. R. (1973). *Aust. J. Agric. Res.* **24**:317)

These findings have important implications for the grazing management of the low growing *S. humilis* and for the selection of companion grasses with a suitable structure. However, some experience in the Northern Territory of Australia suggests that *S. humilis* grows well when lightly shaded by tree canopies. This effect appears to operate through reducing the amount of heat received by the pasture and thus extending the period of moisture availability. However, surface rooted tree species are expected to reduce *S. humilis* growth by competing for water.

(iii) **Temperature**

One controlled environment experiment recorded 33° day/28°C night as the optimum temperature for *S. humilis* vegetative, seedling growth. *S. humilis* was particularly responsive to rises in temperature. Over the range 15°/10°C to 27°/22°C an increase in mean temperature of 2·7°C doubled growth rate of *S. humilis*, whilst an increase of 4·1°C was required to double the growth rate of *Macroptilium atropurpureum*.

Seed production is of particular importance to the continued regeneration of an annual. Some data from University of Queensland research suggest that the adverse effect of cool nights on seed formation is one factor limiting the distribution of *S. humilis* to hot climates. Figure 3.36c shows that seed yield per inflorescence was maximal at night temperatures of 24°C. The yield was unaffected by 31°C day temperature, but decreased steeply at low night temperatures. A companion field experiment showed that the percentage floral abortion was greatest at low night temperatures, no seeds forming when night temperatures fell to a mean of 12°C. Since the number of inflorescences produced (Figure 3.36b) increased with temperature, the total seed yield per plant (Figure 3.36a) was very much higher under hot conditions.

(iv) **Moisture**

Annual legumes may be grown in drier climates than perennial legumes, providing the growing season is sufficiently long and sufficiently reliable to complete seeding in this period. *S. humilis* has regenerated successfully in tropical regions with an annual rainfall of 550 mm and a four-months growing season. On the other hand, annuals are at a distinct disadvantage in climates lacking defined wet and dry seasons.

S. humilis has good seedling establishment characteristics, and will succeed where companion grasses fail because of low rainfall at sowing. However, when growing with an established perennial grass, the percentage botanical composition of *S. humilis* is reduced in subnormal dry years, since the root system of the established perennial grass has a greater share of the available soil moisture. The disease, *Colletotrichum*, is worse in wet years.

(e) Edaphic adaptation

S. humilis is well adapted to soils of low pH containing excess aluminium and manganese. It has a high calcium content on acid soils and will nodulate at pH 4·5, or even at pH 4·0 if the calcium supply is adequate. *S. humilis* has a superior capacity for extracting phosphorus from low-phosphorus soils. However, in Australia, phosphate application has been shown to give a great increase in yield and to be economically worthwhile. These and related questions are explored in Part 5.

S. humilis has shown best performance on light textured, low fertility soils where the competitive strength of companion grasses is low. It does not regenerate well on many cracking-clay soils, and is generally unsuited to saline areas and to soils subject to waterlogging. *M. uniflorum* is similarly best suited to well-drained sites. *M. phaseoloides* is often grown in waterlogged situations with *Brachiaria mutica*. It has a reputation for better performance on heavy soils. This may be because it is easily damaged by nematodes which are present in greater numbers in light soils.

6 Annual fodder crops for the tropics

The requirements for successful choice of annual fodder crops are less exacting than those of perennial pasture species or of grain crops. Cultivation and crop husbandry practices can change the moisture and nutrient availability of the environment, and the competitive position of the sown crop is more favourable than for perennial species. The growth of the crop is usually restricted to part of the year, and hence less tolerance of environmental extremes is needed. Completion of the developmental phase, on which grain crop performance is assessed, is not necessary for fodder crops. They may be grazed off even though environmental conditions may have reduced growth and development. However, considerable scope exists for selec-

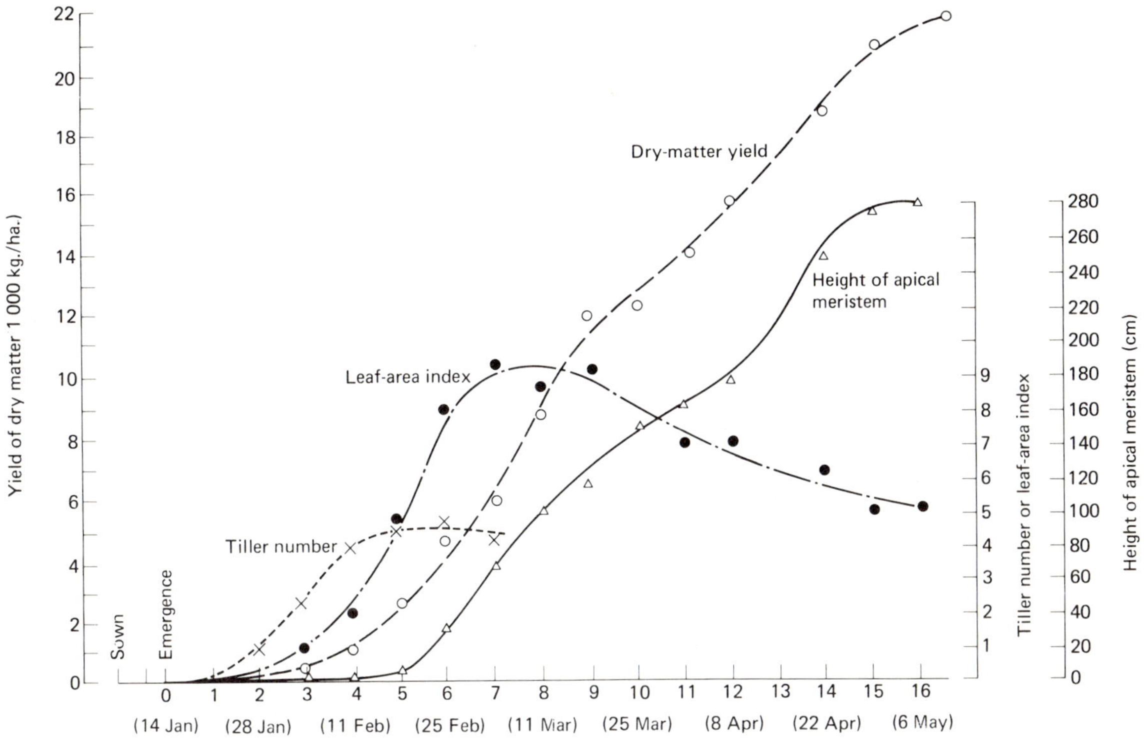

fig. 3.37 Seasonal changes in yield of dry matter, height of apical meristem, leaf area index and number of tillers per plant for *Pennisetum americanum*. (*Source:* Begg, J.E. (1965). *J. Agric Sci.* **65**:341)

ting forage crops best suited to district climatic and edaphic conditions, and to specific animal requirements.

The reasons for sowing annual fodder crops in preference to perennial pastures are related to their superior short-term yielding characteristics, their use in overcoming seasonal feed shortages, their convenience in integrated crop and livestock enterprises, their role in pioneer cropping for land development, and their fodder conservation characteristics. Crops grown primarily for grain are discussed in two companion volumes in this series: 'Tree and field crops of the drier regions of the tropics' and 'Tree and field crops of the wetter regions of the tropics'.

(a) Fodder yield

The highest short-term plant growth rates reported in the world's scientific literature are from annual tropical grass fodder crops. One crop of *Pennisetum americanum* at Katherine, Australia (lat. 14°S) grew at the rate of 540 kg dry matter per ha per day (480 lb per ac per day) for 14 days. This represented the very high conversion rate of 9·5 per cent of the visible radiation received by this rain-fed crop, and occurred after floral initiation when the internodes were lengthening. *Sorghum sudanense* in California (lat. 39°N) grew at 510 kg per ha per day for 35 days, and similar values for sorghum growth have occurred at the University of Queensland's Redland Bay Farm (lat. 28°S). Very high growth rates have also been reported for *Zea mays*, a crop sometimes used for silage.

These very superior growth rates occur for a relatively short period in the crop's development, and are difficult to make use of in animal production. However, high growth rates may be maintained for many weeks after the crop canopy has grown sufficiently to intercept all light. Figure 3.37 shows the course of growth in another *Pennisetum americanum* crop at Katherine. Initial growth rates were low, but in the 5th week after emergence the crop intercepted nearly all the sunlight and high growth rate continued until about the 15th week. Tillering ceased early, but the leaf area index continued to rise to 9 until the 7th week, and decreased thereafter. The final crop yield was 21 700 kg dry matter per ha. The height of the apical meristem rose to 280 cm. The elevated meristem of this crop reduces its regrowth after grazing. The average performance of this crop over 11 years at this site was 13 000 kg dry matter per ha (11 600 lb per ac.).

The best annual fodder crop yield noted in the world literature is for a forage sorghum grown under irrigation at Kununurra, Western Australia (lat. 15°S). This entailed six harvests at the flower-dough stage over 360 days, which gave a total yield of 46 300 kg per ha (41 400 lb per ac.). Higher annual yields have been recorded for perennial crops such as *Pennisetum purpureum* and *Saccharum officinarum.* In southern Queensland at Gatton (lat. 28°S), yields of superior forage sorghums, e.g. *S. almum* cv. Crooble and the hybrid sorghum cv. Krish, have been of the order of 15 000 to 19 000 kg per ha. The development of grain sorghum × *Sorghum sudanense* hybrids has led to substantial increases in first-year dry matter production.

These figures are quoted to indicate the extent to which annual fodder crop yield potential has been realised under experimental conditions. Under farm conditions lower yields are recorded, depending upon the skill of the farmer, the climatic conditions experienced and the soil fertility. Suitable fodder crop varieties have been selected for most types of tropical and sub-tropical environment. A knowledge of genotypic response to climatic conditions is necessary if the best varieties for particular districts are to be identified.

Table 3.16 illustrates how crop behaviour is very sensitive to changes in daylength and temperature. These data are drawn from a controlled environment experiment with inbred lines of *Pennisetum americanum* from the Coastal Plains Experiment Station at Tifton, Georgia (lat. 32°N), and an early (no. 31) and a midseason (no. 23) line have been chosen to illustrate the responses. The flowering of *P. americanum* is accelerated by short days; for instance line 31 grown at temperatures of 28°C day/20°C night flowered in 52 days under 12-hour days and in 73 days under 16-hour days. The difference in flowering time between the two varieties was much greater under long days than under short days. The effect of temperature on flowering was not consistent. The yield of top growth was measured in each treatment ten days after first anthesis occurred, i.e. the slower flowering treatments had a longer growth period before harvest. Table 3.16 shows that yield was very much higher in long-day treatments, which delayed flowering; for instance, line 23 at 28/20°C yielded 48 g in 12-hour days and 217 g in 16-hour days. Growth rates were much higher at mild temperatures under long days; for instance under 16 hour days line 23 yielded 217 g at 28/20°C and 100 g at 36/28°C. The high growth rates exhibited under long days and mild temperatures by line 23 indicates its suitability for use in warm temperate regions.

At Tifton, Georgia the variety Gahi-1 has been very successful, and cv. Tiflate is a later maturing variety suitable for longer season growing conditions. A relative of Gahi-1, cv. Tamworth, is an early variety developed also for warm temperate conditions on the margin of the tropics in northern New South Wales. In the more truly tropical conditions of Northern Australia, cv. Katherine Pearl (ex.

Table 3.16 Daylength and temperature effects on flowering and yield of *Pennisetum americanum* genotypes. (*Source:* Begg, J. E. and Burton, G. W. (1971). *Crop. Sci.* **11**:803.)

Daylength (hours)	**Temperature** day/night (°C)	**Days to First Anthesis**		**Top Growth** (g)	
		Early 31	***Midseason 23***	***Early 31***	***Midseason 23***
12	28/20	52	55	31	48
	36/28	51	56	30	47
14	28/20	64	97	45	231
	36/28	54	82	30	103
16	28/20	73	95	52	217
	36/28	61	100	33	100

Table 3.17 Nitrogen fertiliser level and yield characteristics of *Pennisetum americanum* cv. Gahi—1. (*Source:* Hart, R. H. and Burton, G. W. (1965). *Agron. J.* **57**:376.)

Character	Nitrogen Fertiliser Level (kg/ha)				
	0	*100*	*200*	*400*	*600*
Yield (kg/ha)	5 700	8 700	11 800	14 600	16 000
Yield response (kg DM/kg Nitrogen)		30	30	22	17
Per cent leaf	91	88	85	86	83
Per cent crude protein	8·4	7·8	8·9	12·4	12·7
Per cent crude fibre	28·6	28·6	28·7	28·6	29·4
1 000 IU Vitamin A equivalent per kg	198	178	218	253	275

NB 1 lb per ac = 1·12 kg per ha

Ghana) and the slightly earlier variety cv. Ingrid Pearl (ex. Senegal) have been the best producing varieties.

High fodder yields depend on satisfactory soil fertility. Table 3.17 shows the yield response of *P. americanum* cv. Gahi to increasing levels of nitrogenous fertiliser; this experiment was carried out at Tifton, Georgia (lat. 32°N). Crop yield at this site varied from 5 700 kg per ha in the control plots to 16 000 kg per ha in the plots receiving the very high dressing of 600 kg nitrogen per ha as ammonium nitrate. The data show a typical curvilinear response to increasing fertiliser inputs: at the 100 kg level, 30 kg additional dry matter was grown for each kg of nitrogen applied, whilst the response to 600 kg nitrogen was 17 kg additional dry matter per kg nitrogen applied. Per cent leaf was slightly less in the treatments receiving nitrogen fertiliser, but per cent fibre was unaffected. Per cent crude protein and vitamin A content were higher in the plants receiving very high fertiliser inputs (400 and 600 kg nitrogen per ha), but were not affected by the more moderate fertiliser dressings usually applied by farmers.

Annual fodder crops will often give higher yields than perennial pastures in their first year of establishment. The larger seeded sorghums have higher growth rates and establish more reliably than small seeded perennial grasses such as *Panicum maximum*. Grazing millets, e.g. *Echinochloa crusgalli* var. *frumentacea* or *E. crusgalli* var. *edulis* exhibit good seedling vigour in most farm situations. However, in subsequent years the production of well-adapted, well-fertilised perennial pastures will usually exceed that of annual fodder crops.

(b) Seasonal fodder availability

Annual fodder crops may be used to meet seasonal fodder deficiencies. The pattern of seasonal moisture availability may be radically altered by cultivation practice. Fallowing, which eliminates pasture or weed transpiration losses during the growing season, will conserve water in soils of suitable storage capacity. This may then be used to grow a crop when soils bearing pasture are moisture deficient.

In sub-tropical and high altitude tropical districts the dry season may be sufficiently cool to grow temperate-type fodder crops such as oats (*Avena* spp.) and brassicas. For instance, in sub-coastal southern Queensland, oats are sown after a short summer fallow on the first autumn rains, to provide grazing during the May to October period. This practice is most successful on deep soils which can accumulate summer moisture, and is popular on chernozem soils. Production from beef cattle grazing these crops is usually of the order of 300 kg liveweight gain per ha, with daily rates of gain about 0·8–1·0 kg per head.

The nutritive value of oats is superior to that of tropical grazing crops, and this is reflected in higher daily rates of gain. One series of crops at the Brian Pastures Research Station, Gayndah, Queensland (lat. 26°S) was stocked at 3·6 to 4·5 weaner cattle per ha, and provided 88 to 100 days grazing during June to October, for production varying from 190 to 368 kg liveweight gain per ha, depending on seasonal conditions. At high stocking rates the period of crop availability is often shorter, but similar liveweight gains per unit area are often recorded.

As an alternative to fallowed seed bed preparation, oats may be sod-seeded into existing pasture. This practice is only successful if adequate rainfall occurs during the crop growing period, or if irrigation is available. It usually requires heavy applications of nitrogen fertiliser, since the mineralisation of soil nitrogen which occurs in a fallow is not available in the pasture sod.

Considerable diversity of genetic material is available within *Avena* spp. *A. strigosa* cv. Saia was introduced from Brazil to southern Queensland, where it became very popular in the more humid coastal districts because of its former degree of resistance to stem rust (*Puccinia graminis*) and to crown rust (*P. coronata*). Differences in variety maturity may be used to provide a long sequence of forage availability: early and late varieties may be sown on the one planting rain. Alternatively, late season varieties may be sown on early rains, and quick maturing varieties sown to take advantage of late rains. The correct flowering time of a variety can also be selected on the basis of moisture availability. In Queensland, the *A. sativa* early varieties Bovah and Benton are erect and well accepted by cattle. The mid-season variety Camelia has gained popularity. The late variety Algerian is slower to develop, but has a more decumbent habit with high tiller density, which confers good resistance to grazing. Minhafer and Landhafer varieties are also late maturing, and have outyielded cv. Algerian in many situations. Similar varietal ranges are available in most countries. Barley (*Hordeum vulgare*) is occasionally used for autumn grazing, before oat crops are ready for grazing.

Tap-rooted crops have an advantage in districts where the growth of the crop is almost wholly dependent upon stored soil moisture. Oats may be sown on a single planting rain, or may be sown at depth in stored moisture. The development of crown roots on this plant is essential to provide anchorage, and the seminal root system is unable to exploit the soil mass effectively. Crown root development is conditional upon adequate rains after sowing, and the absence of these will delay readiness for grazing. This has promoted an interest in rape (*Brassica napus* var. *biennis*) and safflower (*Cartharmus tinctorius*) for cool season grazing, since both these crops are tap-rooted. In Queensland, rape is particularly easily damaged by aphid attack. Both crops are less acceptable to stock than oats. Management problems arise with the spiny safflower, and effective use of the crop is difficult to achieve late in the season when the spines are well developed and fibrous. However, the inflorescences are eaten at this stage. Grazing experiments at the Queensland Department of Primary Industries Research Station at Biloela (lat. 24°S) showed similar cattle performance, crop yield and nutritive value of safflower and oats.

Summer grown fodder crops vary in their degree of acceptance by animals grazing them in autumn and winter. The limited usefulness of *Pennisetum americanum* for cool season grazing is discussed in Part 6. Many leguminous crops can be used effectively as standover grazing for use in conjunction with poorer quality cereal residues. Late flowering varieties of *Lablab purpureus* such as cv. Rongai provide good quality cool season grazing when earlier maturing crops have disintegrated. For instance, the varieties of *Vigna sinensis* used in Australia are all much earlier maturing than *Lablab purpureus* cv.

Table 3.18 Yield and nutritive value of *Lablab purpureus* and *Vigna sinensis*. (*Source:* Milford, R. and Minson, D. J. (1968). *Aust. J. Exp. Agric. Anim. Husb.* **8**:409.)

Character	Species	Date of Harvest			
		April 4	*May 15*	*July 11*	*August 15*
Yield (kg/ha)	*L. purpureus*	3 150	5 050	4 700	4 150
	V. sinensis	2 500	2 350		
Per cent leaf	*L. purpureus*	32	31	20	10
	V. sinensis	33	14		
Per cent nitrogen	*L. purpureus*	2·7	2·7	1·9	1·6
	V. sinensis	3·3	2·0		
Dry matter digestibility (per cent)	*L. purpureus*	59·2	57·5	56·6	50·9
	V. sinensis	63·8	59·1		
Dry matter intake (g/kg$W^{0·75}$)	*L. purpureus*	63·8	66·2	73·6	64·1
	V. sinensis	76·0	67·9		

Rongai, which can be used as a companion crop for late maturing maize or sorghum, or may be grazed during the cool season.

Table 3.18 gives data for the comparative performance of these two crops at Gatton, Queensland (lat. 28°S), a district experiencing severe winter frosts. *L. purpureus* gave higher yields than *V. sinensis*, and the crop yield present towards the end of the winter in mid-August was more than 80 per cent of the peak crop yield in mid-May. Some *L. purpureus* deterioration occurred, the percentage leaf decreasing from 32 to 10 and the per cent nitrogen decreasing from 2·7 to 1·6. Although dry matter digestibility fell from 59 to 51 per cent, a very pleasing feature of the results was the voluntary intake (as assessed by penned sheep) was quite well maintained in the standing crop in July and August. *Cajanus cajan* (pigeon pea) is also successfully used for winter grazing.

Where irrigation is available, crops may be grown in the dry season which have better nutritive value than rain-fed crops often grown under lower radiation conditions. Table 3.19 contrasts the chemical composition of four crops of *Zea mays* grown in different seasons of the year at Chiangmai, Thailand (lat. 19°N). All crops were assessed at the same developmental phase, when the grain was at milk stage. Digestibility of the crop harvested in August, in the middle of the rainy season, was 56 per cent, compared with 62 per cent for the crop harvested in January, during the dry season. The latter crop was also higher in protein and lower in fibre.

(c) Integration of crop and livestock production

In some farm situations it is more convenient to integrate cash crop and livestock production by using annual fodder crops in the farm system than by using perennial pastures.

In areas unfenced or with only rudimentary fencing subdivision, the crop rotation programme is easily managed if fodder crops are grown in the situations available for them each season. Perennial pastures require strong fencing, which may be unsuited to the social situation or inappropriate to the investment return from small areas, or strict stock-herding. Annual fodder crops of erect habit are better suited to a single cut and carry operation than to grazing.

When the population pressure on land is heavy, the greater proportion of the farming area will be devoted to food crops for human consumption or for high-value cash crops. Fodder crops may only be grown for part of the year, and multiple-cropping, where a sequence of overlapping crops is grown on the same area, is increasing in favour. In Part 4 the companion cropping of *Stylosanthes guyanensis* and upland cash crops is discussed. This technique corrects protein deficiency in the dry season, and leads to more complete use of low-protein, high-fibre crop stubbles.

Table 3.3 illustrated the benefits of adding a legume to a low quality roughage diet; protein deficiency was rectified, and the correct balance of legume and roughage led to increased animal intake of roughage. Crop stubbles are an important source of animal fodder in many countries. Rice

Table 3.19 Seasonal effects on composition of *Zea mays* at Chiangmai, Thailand (lat. 19°N). (*Source:* Holm, J. (1973). *Thai J. Agric. Sci.* **6**:89.)

Character (as per cent dry matter)	**Harvest Time**			
	August	*November*	*January*	*March*
Crude protein	7·5	8·5	11·3	7·9
Crude fat	2·7	2·9	2·6	2·0
Crude fibre	33·8	27·5	27·9	28·6
Nitrogen-free extract	50·6	54·1	49·8	53·4
Ash	5·3	7·0	8·4	8·0
Digestibility	55·5	57·5	62·1	59·3

crop residues in Asia and sorghum (guinea corn) stubbles in West Africa and Northern Australia are incompletely used, since the protein content may be only 2 to 3 per cent. Companion sown legume crops grazed or cut and fed together with these residues will maintain animals in good condition during the dry season.

Leguminous fodder crops offer a means of maintaining the fertility of soils growing crops which remove a lot of nutrients such as kenaf (*Hibiscus cannabinus*) or cereals, especially if the crops are grazed where they are grown or if animal excreta are carefully returned to the fields. This question is discussed in Part 5. Fodder crops may also be required to provide a break in the crop rotation to minimise a particular weed, disease or pest problem.

Animals are valuable for utilising crops which do not reach maturity. In regions of uncertain rainfall, a simple cash cropping enterprise may be uneconomic; a mixed farming system involving ruminants may be profitable, since cash or grain crop failures may be converted to saleable animal products. Some districts have a rainfall pattern especially suited to an integrated grain/livestock operation. For instance, the Darling Downs district of southern Queensland has highly reliable cool-season cereal grain production. The crops are partly grown on warm-season rainfall stored in these deep soils, and the frequency of gross crop failure has been estimated as one year in ten. On the other hand, summer grain production is less reliable, with a frequency of crop failure estimated as at least one year in five. It is argued that grazed summer fodder crops and winter grain crops provide a farm use pattern better adapted to the environment than either a pure cash cropping or a pure livestock enterprise.

Finally, in regions with a long growing season it is feasible to graze some crops grown primarily for grain, without decreasing grain production. Careful grazing control is required if grain production is not to suffer. Examples of crops used successfully in this way are the mini-milo varieties of sorghum, safflower (*Carthamus tinctorius*), late-maturing oats, and the wheat varieties Hopps and Festiguay.

(d) Pioneer cropping for land development

Virgin land being developed either for permanent pasture or for cash crops (including tree crops such as oil palm and rubber) may benefit from a pioneering fodder cropping phase when the land is first brought under the plough. This is appropriate for a number of development situations.

On low fertility soils, fodder cropping provides a rapid method of organic matter addition, which may be necessary to increase the exchange complex on which soil nutrient retention depends. The use of pioneer leguminous crops such as *Vigna sinensis* or *Stizolobium deeringianum* will also increase soil nitrogen content after only one season's growth, since these nodulate quickly and develop a large mass of active nodule tissue. These crops do not usually benefit from inoculation with *Rhizobium*.

Fodder cropping does not need a great deal of land preparation, since crops may be selected with good seedling emergence and vigour, and these are harvested by grazing stock. This means that the processes of land levelling, final tree stump removal, burning of tree residues, collection of large stems and removal of suckering vegetation may be carried out during the culture and growth of perhaps two forage crops before more demanding cash crops, e.g. irrigated tobacco, are grown.

Fodder cropping provides opportunity for the removal of native species of plants. Experiments at the Brian Pastures Research Station, Gayndah, Queensland (lat. 26°S) showed that cultivation of native bunch grass pastures for six months was inadequate to destroy all native grass and forb seed. Pasture of small-seeded grasses (*Chloris gayana, Panicum maximum* var. *trichoglume* cv. Petrie, or *Cenchrus ciliaris*) with *Medicago sativa* sown after this form of land preparation suffered from native grass competition. However, when a sequence of fodder crops (*Vigna sinensis* with forage sorghum, followed by oats) were sown before the perennial pasture, much more successful sown grass pastures resulted. *M. sativa* established equally well with either native grasses or sown grasses. In managing fodder crops, it is important to prevent seeding down of volunteer weeds, or a fodder cropping phase may cause a greater weed problem for later crops.

When slow-developing perennial pastures are sown, a preliminary fodder cropping programme will bring extra money into the land development enterprise. As indicated earlier, annual fodder will yield more grazing in the first year than small-seeded perennial grasses. The grazing provided in the first year by vigorous, easily established sorghums, perhaps with *Vigna sinensis* or *Lablab purpureus*, will help to pay for land clearing, initial ploughing, fencing and stock-watering.

(e) Conservation of fodder

Growing annual fodder crops which produce a high yield of product suitable for conservation is preferable to cutting lower yields of perennial tropical pasture which are not suitable. As discussed in Part 6, many perennial tropical pasture species are low in sugar, high in fibre, and difficult to compress. *Zea mays* is renowned as a silage crop, and good levels of animal production are assured from this product if it is well made. In less reliable climates, the drought resistant forage sorghums will always provide a harvestable, high sugar silage crop. The juicy, succulent stems of silage varieties such as *S. vulgare* cv. Sugardrip also provide better standover grazing than other grass crops. The content of cynanogenetic glycoside in the sorghums can cause stock poisoning, especially if the crop is used when young or following stress periods. This will not happen if the crop is conserved. The finer stemmed *S. sudanense* gives good yields of hay, and leguminous hays, e.g. from *V. sinensis*, provide higher protein contents which can be used to supplement roughage diets.

Conservation of annual fodder crops has a place in high-farming, intensive enterprises. It is less suited to extensive forms of husbandry where other lower cost inputs, e.g. pasture legume introduction, are more important.

7 Temperate pasture species for cooler tropical situations

Temperate-type pasture species may be useful in the cooler tropical situations, where either high altitude or high latitude modify the climate to make the growing of clovers and rye grasses possible. Irrigation may be needed in regions with a cool dry season, and *Trifolium repens* has formed the basis of many irrigated pastures in sub-tropical districts where whole milk producers needed to maintain continuity of supply through the cooler months.

There are two main reasons for growing temperate pastures in the tropics. As indicated earlier, plants such as *T. repens* and *Lotus major* have an optimum temperature for growth of about 24°C, and will grow at temperatures lower than this where tropical plants such as *Centrosema pubescens* become quiescent. *Lolium perenne* is adapted to cooler temperatures than *T. repens* which has a wide naturalised distribution in the sub-tropics. This pattern of cool season growth is very different to that of most tropical plants, and can be used to overcome seasonal feed shortages.

The second reason is the superior nutritive value of temperate species. *T. repens* has consistently the highest nutritive value of all pasture species, and experimental work in New Zealand has shown that individual animal liveweight gains are higher on *T. repens* than on other diets. Digestibility is usually in the range of 85 to 90 per cent, (compare Table 3.2), with high levels of voluntary intake. Similarly, temperate grasses maintain much higher levels of milk production than are feasible on tropical grasses. This is partly due to the effect of environment on nutritive value; more fibrous feeds are produced in warm conditions, where the plant requirement for structural conducting vessels is greater. However, there are intrinsic genetic differences between temperate and tropical species which are shown in differences in nutritive value when these are grown in the same environment. Dairy producers in the sub-tropics are discovering that cool season grazing on *T. repens, Lolium spp., Avena spp* or *Bromus unioloides* not only maintains milk production when these feeds are available, but builds up body reserves which are partly used for milk production during the warm season. Analysis of the fat constituents in milk can be used to assess the origin of the energy present. This analysis indicates that some of the warm-season milk production from cows grazing tropical pastures comes from body reserves developed earlier in the year on temperate pastures.

There are many difficulties involved in growing temperate species in a tropical environment. The principal problem is the faster growth rate of tropical grasses, discussed earlier in this chapter. This makes temperate swards highly vulnerable to invasion by local volunteer grasses, which are ecologically better adapted to tropical conditions and have a competitive advantage over temperate species, especially during the warm season. In these circumstances it is best to view the sowing of temperate grasses as a temporary measure requiring annual or biennial repetition. Temperate species are also more vulnerable to pests and disease in a warm environment, and their nutritional requirements are often more difficult to satisfy than tropical species.

(a) Temperate grasses

Species of *Lolium* have the highest nutritive value of temperate grasses, and the farmer's choice should always be to grow *Lolium* wherever climatic, soil and management conditions make this possible. Most *Loliums* are large seeded with excellent establishment characteristics, which makes them suitable for short-term pasture production with minimal seed-bed preparation. *L. perenne* is grown under irrigation by producers hoping for persistence. The inter-generic cross *L. perenne* × *L. multiflorum* (H1 rye grass) has better first-year vigour and cool season performance; it is more suited to grazing than many of the annual or biennial *L. multiflorum* varieties. The annual *L. rigidum* is rarely used in the tropics. The large seeded biennial *Bromus unioloides* is used in a similar way to *Lolium*.

Loliums generally demand better moisture and fertility conditions than other temperate species. *Phalaris arundinacea* has more warm season production than *Lolium* and is more tolerant of poor drainage. The sterile hybrid *P. arundinacea* × *P. tuberosa* (Ronpha grass) persists well in irrigated sub-tropical pastures but is poorly accepted by stock. *P. tuberosa* is adapted to heavy soils. It survives hot, dry summers well, since it enters a dormant phase; however this Mediterranean adaptation is out of phase with tropical situations, where warm season rains are usual. *Dactylis glomerata* produces more summer growth than *P. tuberosa*, and is suited to sandier soils. *Festuca arundinaecea* has persisted well in some irrigated sub-tropical situations, but its nutritive value is in the low range for a temperate pasture grass.

(b) Temperate legumes

As mentioned earlier in the chapter, *Trifolium repens* has a creeping habit which makes it more persistent under heavy grazing than many of the commercially used tropical legumes. Some of the large leaved, long-petiole Ladino types have better warm season production than ecotypes from colder regions. In the tropics and sub-tropics *T. repens* sometimes behaves as an annual, especially if high summer temperatures and temporary water shortage expose the stolons to heat stress. In these circumstances selection for good seed production is desirable; Louisiana material seeds down well in the early summer and re-establishes as volunteer seedlings when conditions for white clover growth are more appropriate. Some local ecotypes have been selected for their adaptation to tropical conditions. *T. semi-pilosum* cv. Safari (Kenya white clover) has been released for its superior persistence in coastal sub-tropical districts, and has similar growth characteristics to *T. repens*. Other *Trifolium* spp. are native to the central African highlands.

T. repens is the first species to consider; other temperate legumes are appropriate only if *T. repens* performance does not meet the desired criteria. *Lotus major* and *T. fragiferum* will grow in more poorly drained situations. *T pratense,* which is rather short lived, provides quick feed and better summer production than many clovers. *Medicago sativa* is adapted to dry and well drained situations. Annual larger seeded legumes are suited to early first-year feed requirements and minimal seed bed preparation. Many of these legumes require sowing each year if a cool-season feed deficiency is to be corrected. *T. subterraneum* flowers in response to lengthening days after a period of exposure to cold; seedlings which germinate during the summer will often encounter some stress which causes death before conditions which stimulate flowering occur. Varieties such as Clare or Yarloop, which flower in warmer conditions than other *T. subterraneum* material, are more likely to produce seed for regeneration. *T. alexandrinum* is overseeded in short sod pastures for winter production in Florida and is widely grown in Western Asia and India as a fodder crop. Annual *Medicago* spp. are naturalised on natural and alkaline soils in the sub-tropics; *M. truncatula, M. polymorpha* and *M. littoralis* contribute to cool season protein and forage needs in years when cool season rains occur. Grazing management which prevents accumulation of surplus grass in late summer allows the autumn regeneration of these annuals.

Managerial requirements for the successful use of temperate species in the tropics are much more exacting than for tropical pasture species. In seeking to overcome cool season forage shortages, producers should not overlook plants capable of producing in both hot summers and cool winters; species such as *Setaria anceps, Pennisetum clandestinum, Lotononis bainesii,* and *Desmodium intortum* may have the necessary flexibility.

We should recognise that *P. clandestinum* and *S. anceps* are indigenous to tropical montane areas, where they have a long record of proven superior performance. It is much harder to grow and maintain European plants in the tropics.

8 Tropical pasture seed production

The use in farm practice of the improved plants discussed in this chapter depends upon the continual availability of good quality seed or planting material at reasonable prices. In many tropical countries seed supply has been the main stumbling block to pasture development. Seed of some species, e.g. *Panicum maximum* is chronically in short supply; seed quality of some lines, e.g. *Brachiaria decumbens* has often been poor; prices of species with special harvest difficulties, e.g. *Lotononis bainesii* and *Pennisetum clandestinum,* are high. On the other hand, considerable advance has been made in the last decade in improving techniques of tropical pasture seed production. There have also been advances in identifying the higher seed-yielding varieties, and in developing the grower skills, commercial enterprise and quality control upon which a more efficient industry may be built.

Seed production is the final stage of a sequence of plant processes, some of which can be more easily controlled by the grower than others. The potential yield of the crop is limited by the density of flowers produced; this depends on the degree of plant branching to form shoots, (which is very sensitive to moisture and nutrient supply), the proportion of the surviving shoots which are fertile, and the extent of inflorescence differentiation to form flowers. The yield potential of grass crops is determined before a grower actually sees an inflorescence grow out from a leaf sheath. However, the weight of seed that is actually produced depends upon the proportion of flowers setting seeds, (which is very sensitive to radiation level and conditions for pollination), the size to which the seeds grow, and the ease with which the seeds are recovered at harvest. Harvest efficiency of many tropical pasture seed crops is very low indeed, since flowering proceeds over a long period. In addition, seed of many crops drops to the ground soon after it ripens. Thus only a small proportion of the seed produced may be recoverable by harvest at any one time.

These problems may be partly overcome by careful choice of site, good establishment practice and crop husbandry, and by appropriate harvesting and processing techniques.

(a) Choice of site for seed production

Many pasture seed production enterprises have failed because of unsuitable choice of location. High seed yields do not occur in high rainfall, low radiation equatorial regions where flowering is restricted and disease flourishes. Similarly, late autumn-flowering pasture legumes will not reliably produce seed in areas subject to frost. Pasture seed is a high value crop requiring suitable soil types if the necessary high investment in husbandry is to receive adequate return.

The climate of the district selected should provide adequate length of growing season to build a satisfactory seed yield potential, suitable length of day and temperature to promote flowering of the varieties to be multiplied. It should also provide even, sunny conditions during crop maturation to promote flower-opening, pollination of insect-dependent crops, and high rates of photosynthesis during seed differentiation. The availability of irrigation provides good control of crop development, and is often necessary if seed is to be grown in dry districts with high radiation input.

The seed production field must also be suitably isolated from potential contaminants. An isolation distance of 20 m may be quite adequate to prevent mechanical contamination of apomictic species such as *Cenchrus ciliaris.* Cross-pollinated crops such as *Setaria anceps* should be spaced about 100 to 200 m from contaminating varieties of the same species.

(b) Crop establishment

Pasture establishment principles are discussed in Part 4. Special attention to weed control and to control of plant density is necessary for seed production crops. Investment in weed control is also necessary since, in contrast to normal pasture situations, the presence of some weeds may make the seed crop quite unsaleable. Weed seeds which are difficult to remove during seed harvesting and processing are the first consideration; weeds which reduce crop growth through competition are of secondary importance. For instance, *Sida* spp. *Stachytarphaeta urticifolia, S. jamaicensis* and *Crotalaria* spp. are of interest to *Stylosanthes guyanensis* growers, since the processed seeds are difficult to separate with normal cleaning operations. Chemical weed control measures are often profitable for seed crops, where the returns from grazing alone may not justify such expense. Pasture legume seed crops are particularly vulnerable to weed invasion. Careful site selection can prevent weed problems; virgin grassland and some virgin woodland sites cause fewer problems for the seed producer than old cultivation land.

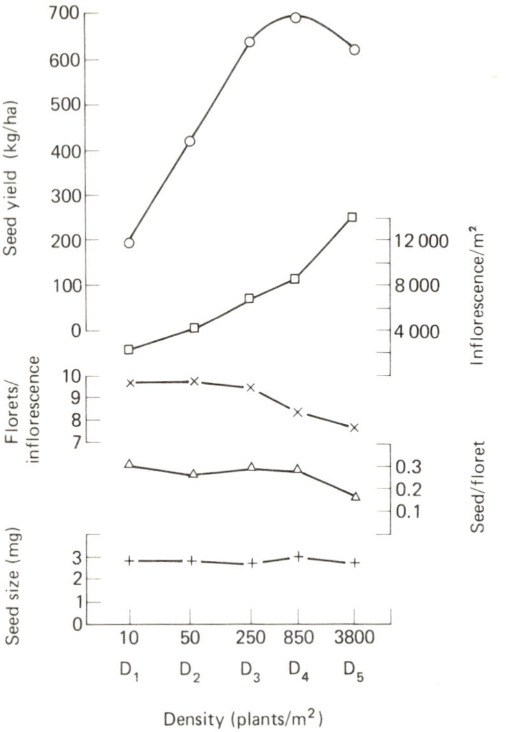

fig. 3.38 Influence of density of *Stylosanthes humilis* on seed yield and its components. (*Source:* Shelton, H. M. and Humphreys, L. R. (1971). *J. Agric. Sci.* Camb. **76**:325)

Table 3.20 Nitrogen supply and row spacing effects of seed production of *Setaria anceps* cv. Nandi II. (*Source:* Boonman, J. G. (1972). *Neth. J. Agric. Sci.* **20**:22.)

Nitrogen Level	**Pure Live Seed Yield** (kg/ha) *Width between Rows* (cm)		
	30	**60**	**90**
Low	7·6	11·7	13·1
Medium	19·5	18·9	14·9
High	19·0	14·0	11·0

NB 1 lb per ac = 1·12 kg per ha

The control of plant density is desirable in annual seed crops. Fig. 3.38 shows the response of irrigated *Stylosanthes humilis* to variation in density. Maximum seed yield was 690 kg per ha at an optimum density of about 850 plants per m^2; seed yield was reduced at both higher and lowest densities. Seed yield was linearly related to the logarithm of plant density up to 250 plants per m^2. This density gave about 90 per cent of maximum yield. Seed producers in the Northern Territory of Australia, for example, sowing *S. humilis* at 25–40 kg seed pods per ha should reach this density in the first year. The effect of plant density on the components of seed yield are shown in Fig. 3.38. The number of inflorescences formed per unit area was the most important factor controlling yield; the low seed yields of low density stands was due to shortage of inflorescences. On the other hand, in the highest density sward increased inflorescence density did not quite compensate for the reduced floret number per inflorescence and the poorer seed set recorded in this treatment. Seed set was generally low in this cool Brisbane (lat. 28°S) environment. Density did not significantly affect seed size.

Perennial grasses may be sown at lower rates than annual legumes, since these branch freely to generate inflorescence bud sites so that the density of sowing becomes less important. It is generally desirable to establish the crop in rows, since this makes weed control easier and provides more reliable conditions for seed production in dry seasons. Table 3.20 shows how the optimum spacing between rows decreases with increase in fertility level. This work was carried out with *Setaria anceps* Nandi II at Kitale, Kenya (lat. 1°N). At low nitrogen level (no fertiliser nitrogen applied) the yield of pure live seed was at a maximum of 13·1 kg per ha at 90 cm row spacing. It was reduced at the closer inter-row spacings of 60 and 30 cm. Conversely at medium nitrogen level (130 kg N per ha per crop), pure live seed yield of 19·5 kg per ha was maximal at 30 cm row spacing, and was reduced at 90 cm inter-row spacing. This reduction was mainly due to decreased fertility of tillers. Seed production of runnering grasses is less sensitive to variation in density than that of bunch-type grasses.

(c) Crop husbandry

High fertiliser-application levels on seed crops are usually profitable and soil nutrient deficiencies should be identified and rectified with a balanced fertiliser programme.

fig. 3.39 A seed crop of *M. atropurpureum* ready for harvest

Particular attention should be given to boron and copper availability, since seed crops are more sensitive to these deficiencies than grazing crops. Grass seed yields are closely related to nitrogen supply, as indicated in Table 3.20, and optimum levels of application for crops which have enough water are usually in the range 100 to 200 kg N per ha per crop. Crops suffer more from lodging at higher levels of nitrogen. Applications at the beginning of seasonal growth are usually more effective than applications late in the crop development phase, but a policy of split applications provides the seed grower with more flexibility in meeting particular seasonal conditions, and is necessary on very light soils subject to leaching.

Incidence of weeds, pests and diseases is local in character, and local control schedules need to be devised. Most pasture managers do not employ chemical control measures against pests and diseases but these may be needed for pasture seed crops. It is important to use only seed which is disease and pest-free. Seed born diseases such as *Psuedomonas phaseolicola* (halo-blight) are dangerous not only for *Macroptilium atropurpureum* cv. Siratro crops but also for neighbouring vegetable crops in related genera. With insect-pollinated crops such as *Desmodium intortum* care must be taken that spray schedules do not damage beneficial insects.

Grazing or cutting pasture seed crops provides an additional income source, and may even increase seed yields. There is also the problem of residue disposal after seed harvest. In some disease situations residues should be burnt; otherwise residues may be sold as fodder or converted to animal product by the producers. To prevent seed contamination from grazing, animals which have been fed on crops that might be contaminated should not be introduced into a seed producing field until after they have voided these crops.

To decide the best plan of grazing management for seed crops, a very careful study of individual crop development is needed, and research guide lines are limited. Defoliation of *Pennisetum clandestinum* is essential to stimulate flowering, since the axillary buds are inhibited by the dominant apical buds unless these are removed. Defoliation delays flowering in some species, and this may be harmful or helpful, depending upon the environmental conditions occurring during the delayed flowering and maturation period. Late grazing which removes the flowering parts is usually harmful, but early grazing may stimulate branching and the generation of more sites for flower

fig. 3.40 Short, hairy pods of the trailing *Calopogonium mucunoides*

fig. 3.41 Windrowing cut material of *M. atropurpureum* before suction harvesting seed from the ground

production, especially if it is not so severe that it decreases sward growth rate.

Careful grazing reduces the risk of lodging in well grown crops, and decreases the amount of leafy material to be handled in the harvester. Crops of twining legumes such as *Glycine wightii, Macroptilium atropurpureum* cv. Siratro and *Macrotyloma axillare* are easier to cut and thresh if grazed, and it is usual to close the field from grazing about 12 weeks before harvest. Rolling with a heavy flat iron roller is an alternative to grazing. This compacts the crop, produces a narrower band of branching sites which produce a more elevated distribution of seed heads, and also reduces weed growth.

Grazing or cutting can be used to favour the competitive position of the sown crop species. Low-growing shade-intolerant legumes such as *Lotononis bainesii, Stylosanthes humilis* and *Trifolium semi-pilosum* are favoured by fairly heavy defoliation which prevents taller grass or weed invaders from shading them.

(d) Harvesting and processing

Optimum harvest time is especially difficult to judge. The hazards of weather are common to all seed producers, but tropical pasture seed producers are usually dealing with crops which produce inflorescences over a long period, and these ripen unevenly and shed seed quickly. The grower tries to judge the stage when the rate of increase of ripe seeds from young inflorescences balances the loss of good seed from older inflorescences. One rule to follow is that the first formed inflorescences produce more, and better quality, larger seeds than later-formed inflorescences. Optimum harvest time usually occurs before maximum inflorescence density occurs. *Setaria anceps* which flowers over a long period is usually harvested five to six weeks after the first head appears, depending on weather conditions.

Table 3.21 summarises the results of a study at Samford, Queensland (lat. 27°S) comparing different methods of harvesting *Panicum coloratum*, a plant with poor harvest characteristics. A collection of nearly all of the seeds was made by suspending hessian slings beneath the plants, and this gave a yield estimate of 410 kg per ha of seed of 54 per cent viability. A single cut followed by field drying gave only 104 kg per ha of lower quality seed, and this yield was similar to that obtained with a direct header harvester. The reaper and binder resulted in substantial seed losses, but hand shaking and collection on seven occasions over 15 days gave a better result. Hand harvesting of tropical pasture seed, particularly if labour is experienced and well supervised, can lead to high yields of very good quality seed.

The bulk of mechanically harvested crops are directly headed; drying is usually necessary after the cutting, threshing, and cleaning operations. *Desmodium intortum* is cut and dried in the field before threshing, to allow further ripening and to aid threshing. Some grass seed crops, e.g. *Cenchrus ciliaris, Paspalum dilatatum*, are harvested on several occasions during the one season using a beating and rubbing principle which shakes the ripe seed from the heads. Finally, some legume seed crops, e.g. *Stylosanthes humilis* and *M. atropurpureum* are sucked or swept up from the ground surface after the mature crop residues have been removed.

Table 3.21 Harvest method and seed yield of *Panicum coloratum*. (*Source:* Roe, R. (1972). *Trop. Grassl.* **6**:113.)

Method	Seed Yield Recovered (kg/ha)	Seed Size (mg)	Germination (per cent)
Total collection	410	0·98	54
Cutting and field drying	104	0·96	39
Hand shaking	256	1·03	53
Reaper and binder	31*	0·76	29
Direct heading	123	0·74	35

* Excluding seed shed, some of which was later recovered

fig. 3.42 A suction harvester and bagged seed of *M. atropurpureum*

Seed cleaning to remove seeds of contaminant species, and to separate high quality seed from soil, chaff, dust and inferior seed relies on may ingenious procedures. These are based on differences in seed size, length and width, weight, specific gravity, surface texture and shape. Storage of grass seed in air-tight sealed containers after drying to 8 to 10 per cent moisture is a necessary protection in tropical conditions. Safe packaging, clear labelling which indicates variety and source of seed, and quality control testing are necessary for good trading relations in tropical pasture seed.

Further reading

Anderson, G. D. (1968). Promising pasture plants for northern Tanzania. *E. Afr. Agric. For. J. 34*:106.

Arora, R. K., and **Chandel, K. P. S.** (1972). Botanical source areas of wild herbage legumes in India. *Trop. Grasslds. 6*:213.

Bogdan, A. V. (1965). Cultivated varieties of tropical and sub-tropical herbage plants in Kenya. *E. Afr. Agric. For J. 30*:330.

Bogdan, A. V., and **Mwakha, E.** (1970). Observations on some grass/legume mixtures under grazing. *E. Afr. Agric. For. J. 36*:35.

Bowden, B. N. (1962). An illustrated field key to the commoner Uganda grasses. *E. Afr. Agric. For. J. 27*:230.

Burt, R. L., Edye, L. A., Williams, W. T., Grof, B., and **Nicholson, C. H. L.** (1971) Numerical analysis of variation patterns in the genus *Stylosanthes* as an aid to plant introduction and assessment. *Aust. J. Agric. Res. 22*:737.

Cameron, D. G. (1973) Lucerne (*Medicago sativa*) as a pasture legume in the Queensland sub tropics. *J. Aust. Inst. Agric. Sci. 39*:98.

Chester, R. D., Marriott, S., and **Harvey, J. M.** (1957). Observations on top-dressing a copper-deficient soil in south-eastern Queensland. *Qld. J. Agric. Sci. 14*:1.

Davies, J. Griffiths, and **Hutton, E. M.** (1970). Tropical and sub-tropical pasture species. (cont. in) *Australian Grasslands.* Aust. Nat. Univ. Canberra.

Ford, C. W. and **Williams, W. T.** (1973) *In vitro* digestibility and carbohydrate composition of *Digitaria decumbens* and *Setaria anceps* grown at different levels of nitrogenous fertiliser. *Aust. J. Agric. Res. 24*:309.

Gates, C. T., Williams, W. T. and **Court, R. D.** (1971). Effect of droughting and chilling on maturation and chemical composition of Townsville stylo (*Stylosanthes humilis*) *Aust. J. Agric. Res. 22*:369.

Hacker, J. B. (1972). Seasonal yield distribution in *Setaria. Aust. J. Exp. Agric. Anim. Husb. 12*:36.

Harding, W. A. T. (1972) The contribution of plant introduction to pasture development in the wet tropics of Queensland. *Trop. Grasslds. 6*:191.

Hendricksen, R. E. (1970) The productivity under grazing of *Avena strigosa* alone and in combination with *Medicago sativa*. Proc. XI Internat. Grassld Congr., Surfers' Paradise, Australia. p. 814.

Hopkinson, D. (1970) The potential of some pasture plants in the sisal areas of the Tanga region of Tanzania. *E. Afr. Agric. For. J. 35*:299.

Humphreys, L. R. (1974) '*Tropical Pasture Seed Production*' FAO, Rome.

Hutton, E. M. (1962). Siratro—a legume bred from *Phaseolus atropurpureus. Aust. J. Exp. Agric. Anim. Husb. 2*:117.

Hutton, E. M. (1970) Tropical Pastures. *Adv. Agron. 22*:1.

Hymowitz, T. (1971) Collection and evaluation of tropical and sub-tropical Brazilian forage legumes. *Trop. Agric. 48*:309.

Imric, B. C. (1973) Variation in *Desmodium intortum*: a preliminary study. *Trop. Grasslands* 7:305.

Keya, N. C. O., Olsen, F. J. and **Holliday, R.** (1971) Oversowing improved pasture legumes in natural grasslands of the medium altitudes of western Kenya. *E. Afr. Agric. For. J. 37*:148.

Lawton, R. M. (1968). The value of browse in the dry tropics. *E. Afr. Agric. For. J. 33*:227.

Lloyd, D. L. (1971) Growth and development of makarikari grasses and three other subtropical species grown in the field at Toowoomba, Queensland. *Aust. J. Exp. Agric. Anim. Husb. 11*:525.

Loomis, R. S. and **Williams, W. A.** (1963) Maximum crop productivity: an estimate. *Crop. Sci. 3*:67.

Ludlow, M. M. and **Wilson, G. L.** (1970) Studies on the productivity of tropical pasture plants II. Growth analysis, photosynthesis, and respiration of 20 species of grasses and legumes in a controlled environment. *Aust. J. Agric. Res. 21*:183.

Nicholas, D. B. and **Haydock, K. P.** (1971) Variation in growth and nodulation of *Glycine wightii. Aust. J. Agric. Res. 22*:223.

Olsen, F. J. and **Tiharuhondi, E. R.** (1972) The productivity and botanical composition of some selected grass/legume pasture mixtures at different seeding rates. *E. Afr. Agric. For. J. 38*:16.

Parbery, D. B. (1966) Multiple harvesting of irrigated sorghum in northern Australia. *J. Aust. Inst. Agric. Sci. 32*:44.

Pritchard, A. J. (1964) Comparative trials with *Sorghum almum* and other forage sorghums in south-east Queensland. *Aust. J. Exp. Agric. Anim. Husb. 4*:6.

Queensland Dept. Primary Industries (1968) Grazing oats in Queensland. *Qld. Agric. J. 94*:72.

Reid, R. (1973) A numerical classification of sown tropical pasture regions based on the performance of sown pasture species. *Trop. Grasslands 7*:331.

Smith, F. W. (1972) Potassium nutrition, ionic relations, and oxalic acid accumulation in three cultivars of *Setaria sphacelata. Aust. J. Agric. Res. 23*:969.

Stobbs, T. H. (1969) The influence of inorganic fertilisers upon the adaptation, persistency and production of grass and grass/legume swards in eastern Uganda. *E. Afr. Agric. For. J. 35*:112.

Stobbs, T. H. (1971) Production and composition of milk from cows grazing siratro (*Phaseolus atropurpureus*) and greenleaf desmodium (*Desmodium intortum*). *Aust. J. Exp. Agric. Anim. Husb. 11*:268.

Taerum, R. (1970) Comparative shoot and root growth studies on six grasses in Kenya. *E. Afr. Agric. For. J. 36*:94.

Thomas, R. and Humphreys, L. R. (1970) Pasture improvement at Na Pheng, central Laos. *Trop. Grasslds 4*:229.

Tow, P. G. and **Walker, R. W.** (1971) Winter growth of a glycine-green panic pasture in a typical upland environment. *Aust. J. Exp. Agric. Anim. Husb. 11*:640

Walker, B. (1969) Herbage plants for pasture improvement in Western Tanzania. *E. Afr. Agric. For. J. 35*:1.

West, S. H. (1971) Biochemical mechanism of photosynthesis and growth depression in *Digitaria decumbens* when exposed to low temperatures. Proc. XI Internat. Grassld Congr. Surfers Paradise, Australia. p. 514.

Whiteman, P. C. (1968) The effects of temperature on the vegetative growth of six tropical legume species. *Aust. J. Exp. Agric. Anim. Husb. 8*:528.

Part 4 Pasture establishment

1 Introduction

Pasture establishment is, in many tropical environments, the most hazardous process in the sequence of events leading to improved pastures since success depends on the weather after planting. The intelligent selection of species well adapted to the planting situation and the production of high quality seed, as discussed in the last chapter, make sowings more reliable. The operator must also understand the physical and biological processes which determine the outcome of field sowings and the options available to him so that his management decisions will be based on a realistic estimate of the likely returns from different intensities of pasture establishment inputs.

The first aim when establishing pasture is to provide an environment which is favourable for seed germination, seedling emergence and growth, or favourable for planted vegetative material to initiate new roots and shoots. Another aim is the destruction or retardation of unwanted plants, which might otherwise dominate the sown species. There are different ways of doing those things depending on the sowing situation. Clearing of virgin land and its seed bed preparation require a different approach from that needed for upgrading pasture lands which have already been changed by previous grazing use and in which weed problems have arisen. When pastures are sown in companion cash crops there are further requirements. Less money may be spent on ensuring reliable and effective pasture establishment in dry or saline environments, where it would take longer to pay for intensive pasture development, since the productivity which the pasture is capable of is lower.

2 Seed quality

Seed quality is defined first by the proportion of seeds which will germinate to form healthy plants and secondly by the freedom of the seed from contamination by seeds of different genetic constitution, by inert material or by pests or diseases. Other considerations such as longevity in storage, rate of germination, vigour, dormancy, origin and size may also be involved.

(a) Seed viability

The pasture seed is a dry, resting structure comprising an embryo and endosperm energy reserves, surrounded by a seed coat or testa and possibly by other outer coverings. When it is placed in a moist environment falling within a specific temperature range, the seed absorbs moisture and various biochemical changes begin. Plant hormones called gibberellins produced by the embryo start changes in the starchy aleurone layer surrounding the embryo. Hydrolytic enzymes break some endosperm reserves down to more mobile forms of soluble carbohydrate and amino acids, and the growing embryo uses the latter in protein

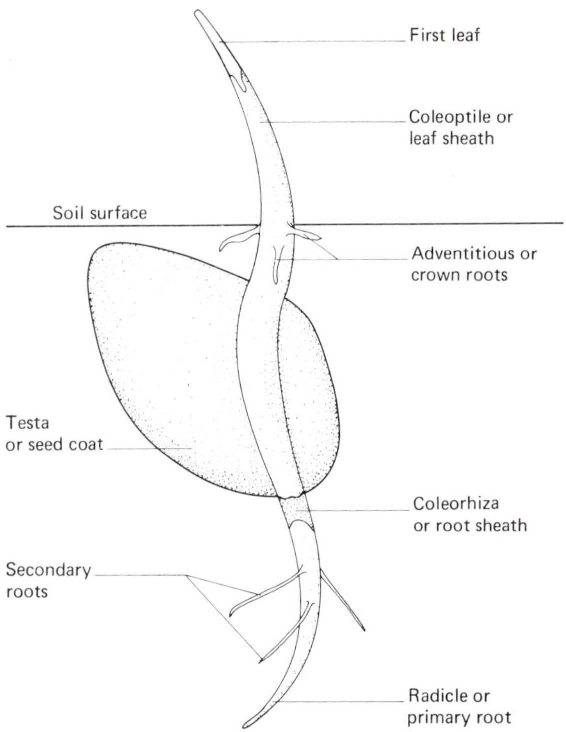

fig. 4.1 Schematic diagram of an emerging grass seedling

synthesis. These activities are also dependent on a favourable supply of oxygen, and carbon dioxide and light may be needed to stimulate the germination of some seeds. As the embryo grows and elongates, the first root or radicle is forced through the coleorhiza or sheath (Fig. 4.1). The bursting of the radicle through the testa marks the first visible germination phase. This is closely followed in grasses by the appearance of the coleoptile with the first foliage leaf enclosed, and in legumes by the two seed leaves or cotyledons. Successful pasture establishment (as distinct from germination) occurs when the radicle becomes anchored in the soil, absorbing moisture and nutrients for the growing shoot. This has emerged through the soil surface to display a leaf surface photosynthesising sufficient energy to make the seedling independent of its seed energy reserves.

Table 4.1 Standards for some certified seed in Queensland. (*Source:* Queensland Department of Primary Industries.)

Parameter (per cent)	***Setaria anceps***	***Stylosanthes humilis***
Germination minimum	20	40
Pure seed minimum	60	90
Hard seeds maximum		20
Inert matter maximum	39	9·5
Weed seed maximum	0·2	0·5
Other crop seed maximum	1·0	0·5

Clearly a high content of dead seed which will not germinate represents a loss to the purchaser, and many countries lay down minimum standards of seed viability which the supplier of the seed is required to meet. An example of these standards is given in Table 4.1. Good seed producers will exceed the minimum viability standards of 20 per cent for *Setaria anceps* and 40 per cent for *Stylosanthes humilis* and be able to command a high price for their seed from educated merchants and farmers.

(b) Freedom from contamination

A notable advance in tropical farming practice is the availability of pasture varieties of predictable performance, seed of which is sold true to the label on the seed lot purchased. Contamination by 'off-types' can occur. For example in a sample of seed labelled *Setaria anceps* cv. Narok seeds of a less frost tolerant cultivar may be found. Such contamination reduces the superiority of the genotype sown and restricts the possibility of using the pasture sown for further seed production. Contamination by other crop species can also cause problems, e.g. *Chloris gayana* seed in a *Panicum maximum* var. *trichoglume* seed lot in Australia, reduces the value of grazing if *P. maximum* is better accepted by stock. Disastrous consequences may occur if the contaminant is a well adapted weed species; e.g. the spiny *Cenchrus echinatus* has been introduced to one location in Laos as an accidental import of foreign pasture seed.

A second form of contamination is the inclusion of inert material—chaff, dust, pieces of straw, and soil in the seed lot. The farmer will often pay high prices per kg of 'seed', much of which does not represent seed at all. The amount of inert material present is measured by the analysis for 'purity'. The high figure in Table 4.1 of 39 per cent inert matter by weight illustrates the difficulty of cleaning some tropical grass seeds, e.g. *Setaria anceps*, which have light spikelets of specific gravity similar to chaff. Sometimes the legal analysis for purity overvalues the seed, since 'seed' may be distinguished as florets, spikelets, or seed units which include glumes and bristles attached to the true seed. The seed units or 'fascicles' of *Cenchrus ciliaris* usually contain 70 to 75 per cent of glumes and bristles, and only 25 to 30 per cent by weight of caryopses. Similarly, seed of *Stylosanthes humilis* is usually sold contained in hooked pods; the pod's outer casing comprises about 35 per cent of the pure 'seed'.

Finally, high quality seed should be free of seed-borne diseases, pieces of sclerotia, insect pupae, etc. The sale of seed of *Macroptilium atropurpureum* cv. Siratro infested with halo blight (*Pseudomonas phaseolicola*) has been a great danger to adjacent producers of horticultural crops.

When determining seeding rates or comparing prices of seed lots it is useful to calculate the pure live seed (PLS) content of a seed lot, which is the product of percentage viability and percentage purity. For example, consider two seed lots of *Paspalum plicatulum*; lot A is of 90 per cent purity and 60 per cent viability and retails at \$2.16 per kg, lot B is of 80 per cent purity and 40 per cent viability and retails at \$1.60 per kg. The price per kg PLS for Lot A is $\$2.16/0{\cdot}9 \times 0{\cdot}6 = \$2.16/0{\cdot}54 = \$4$. The price per kg PLS for Lot B is $= \$1.60/0{\cdot}8 \times 0{\cdot}4 = \$5$, which is more expensive.

(c) Seed longevity

Viability is the first criterion of quality. However, the maintenance of viability for long time periods may be an important consideration for the seed user who stores seed for sowing in a subsequent year (perhaps because good sowing conditions are delayed), or who sows seed some weeks or even months before good germinating rains fall.

In these circumstances dormant seeds will maintain viability; this is discussed later. Tropical legume seeds often store well for this reason. Non-dormant seed has a longevity which differs from seed to seed. The periods of viability of individual seeds are randomly distributed about some average value, so that the longevity of a seed lot is best described in terms of these population characteristics. The death of seeds is increased by high seed moisture content (which is influenced by the relative humidity of the storage atmosphere), by high temperature, and also by a high oxygen content of the storage atmosphere. These relationships may take the form of a sigmoid curve of seed deaths against time, with the average longevity of seed (p_{50} = period for 50 per cent of seed to lose viability) determined as follows:

$$\log p_{50} = K_v - c_1 m - c_2 t$$

where m is the per cent moisture, t is the storage temperature, and K_v, c_1, and c_2 are constants.

The storage of tropical grasses and of soft legume seed constitutes a very real problem. The complete seed population may die within one season of storage in porous sacks in the tropics; this has occurred with *Setaria anceps* and *Chloris gayana*. Seed drying to 8–10 per cent moisture followed by sealed storage in non-porous bags, is the first requirement in the tropics; cold storage is a further refinement. It is generally considered that each reduction of between one and two per cent in seed moisture content doubles storage life; reducing storage temperature 5°C has a similar effect. However, only seed of high initial vigour will keep well.

(d) Rate of seed germination

A rapid rate of seed germination is often beneficial to field establishment. Competition with fast germinating weeds or survival before soil drying are both favoured if the seeds sown germinate quickly. Rate of germination has some mathematical similarities to rate of seed death and may be considered as the probability of a proportion of seeds in a seed lot germinating in a given period of time. It is convenient to separate two aspects of seed quality; the total viability of the seed lot, and the speed with which seeds will germinate. This rate is characteristic of any seed population germinated under standard conditions. Table 4.2 illustrates how the total germination of *Chloris gayana* was very similar over a very wide range of temperature conditions. However, the time for 60 per cent of the viable seeds to germinate varied from 24 hours at 36°/18°C day/night temperature to 164 hours at 18°/5°C.

Table 4.2 Temperature and *Chloris gayana* germination. (*Source:* Leslie, J. K. (1970). Ph.D. Thesis University of Queensland.)

Temperature day/night (°C)	**Germination** (per cent)	**Time to 60% of Maximum Germination** (hours)
36/18	82	24
29/18	78	27
18/15	84	40
18/10	87	92
18/5	79	164

Work by the Queensland Department of Primary Industries at Toowoomba has shown that the rate of germination may be broken up into two independent periods; the latent period from the onset of imbibition to the visible bursting out of the radicle in the first germinating seeds, and the subsequent germinability of the sample. A better understanding of the seed harvest and storage conditions favouring rapid germination would be helpful. Seed testing data which give an indication of the rate of germination will help the seed user buy seed with good establishment prospects.

(e) Seed vigour

Scientists do not all agree on what is meant by 'vigour' or on how it should be measured. Vigorous seed lots germinate readily over a wider range of environmental conditions, establish more reliably in the field, and give higher crop yields. Some merchants in the USA are now giving a vigour rating on their seed labels, based on rate of germination, rate of hypocotyl elongation, and viability in cold soil where pathogens flourish. Staining cut seeds with a chemical called tetrazolium can measure viability, and the depth and location of staining enables a skilled operator to predict seed vigour.

(f) Seed dormancy

Seed dormancy is a natural protective phenomenon which prevents all the seeds of a population from germinating on one occasion. This prevents the loss of all reserves of seed from the soil after a germinating rain. In some environments drought, extremes of temperature, pests or disease may cause the death of all the germinated seeds. Dormancy is therefore an insurance measure; in reliable environments farmers will consider this insurance too costly, and treat seed to remove dormancy and so ensure more complete germination of the seeds sown. This is discussed in a later section. Seed dormancy also assists longevity in storage.

Dormancy depends on a number of factors. If we exclude from discussion seeds which do not germinate because they were harvested before the embryos had developed normally, we may recognise two types of dormancy. First, there is embryo dormancy in which the embryos are physiologically inactive, due to inactive enzyme systems, and second there is dormancy associated with seed coat characteristics. The latter may be due to the seed coat mechanically constricting the expansion and growth of the embryo, as occurs in *Urochloa mosambicensis*, or preventing the entry of moisture or gases, as occurs in many legumes. The seed coat may also contain a chemical which inhibits germination, as has been reported in *Cenchrus ciliaris*. The climatic conditions occurring during seed maturation influence how much hard seed develops, more hard seed being produced in a season with a long, even maturation period. The proportion of hard seed decreases naturally with age, and with exposure to alternating or to high temperatures. In field plots at Townsville (lat. 19°S) a number of lines of *Stylosanthes humilis* had 74 to 99 per cent hard seed in July; by November, seed on the ground was 31 to 49 per cent hard. Table 4.1 suggests that a maximum hard seed content of 20 per cent is standard for seed sold of this self-regenerating annual species.

(g) Seed origin

Identification of the district where the seed was produced can give an indication of quality for some species, if rapid genetic shift is possible or if local adaptation is well developed. For example, a farmer from a cool district would be unwise to buy seed of the frost-tolerant Narok cultivar of *Setaria anceps* if seed of this open pollinated variety had been produced for a long period in the warm tropics. *Stylosanthes humilis* varies greatly in its length of growing season. A farmer living in a short season district receiving about 700 mm annual rainfall should not buy seed of unregistered varieties produced in a long season district receiving about 1 200 mm annual rainfall, since the seed is likely to give rise to late flowering plants which may not set much seed in his district. Conversely a farmer in a wet district loses much production potential by buying early flowering lines from a dry district.

On the other hand, district origin will be of little account when purchasing known varieties of genetically stable plants, e.g., of *Cenchrus ciliaris* with its apomictic breeding system which was discussed in Part 3.

(h) Seed size

Seed size is the most important component of seedling vigour, as indicated in an earlier chapter. Many species produce seedlings whose early relative growth rates are similar. In these circumstances, the seedling which starts with the largest seed produces the biggest plant. Comparisons of tropical pasture species at the University of Queensland have shown that this generalisation was not true for some species, e.g. *Centrosema pubescens, Pueraria phaseoloides*. These species have a lower proportion of their seed as embryo; species with the biggest embryos produce the biggest seedlings. Eventually some other factor, such as inter-plant competition, limits the size of the seedling, but the larger seeds have an early advantage in speed of emergence from depth, root growth, rate of leaf appearance and leaf size; these all help successful pasture establishment.

Individual seed weight varies less than many other cultivar characteristics, but it is common to find seed lots varying plus or minus 30 per cent of their average seed weight, according to conditions of seed production or grading. If seeds are larger then there are fewer per unit weight, but these establish more reliably.

(i) Control of seed quality

In many tropical countries there are some regulations for control of seed quality. The minimal levels of control are phytosanitary procedures to restrict the import of disease and of weeds, and minimum standards of quality for seed offered for sale, which are legally enforceable. These controls protect the seed user and the seed grower or merchant who wishes to maintain sales on the basis of reputation. The regulations depend for their success on the existence of well-run seed testing laboratories who use the standard rules for sampling, purity and viability testing agreed upon by the International Seed Testing Association.

Certified seed represents a more sophisticated development, in which the genetic authenticity of the seed, as well as its viability and purity, is guaranteed by the certifying authority and seed is sold under the official seal. There are various types of organisations certifying seed. In some countries they are grower dominated with government taking an advisory role. Alternatively they may be jointly operated by grower and government, or government run. They involve grower and land registration, inspection during crop production, harvest and processing and laboratory testing. 'Growing-on' tests may also be conducted, in which seed is sown and the resultant plants are examined to check genetic homogeneity. Successful schemes depend upon demand for certified seed and flexible organisational procedures with low overheads. It is at least desirable that sufficient certified pasture seed be available to meet the needs of the seed grower on whom further multiplication depends.

Considerable progress has occurred in the past two decades in improving the standard of tropical pasture seed offered for sale. Seed users need no longer accept inferior seed which can put the success of the pasture establishment operation in danger from the beginning.

3 Factors restricting the success of establishment

So many factors determine whether pasture establishment will be successful or not in different environments that it is difficult for scientists to decide their relative importance. Basically we may be concerned with four types of seed problem. First there is physical loss of seed, perhaps due to it entering the food chain of predator harvesting ants or to its physical movement to inaccessible depths on cracking soils. The second problem is loss of seed viability, which is often associated with seed weathering. A third problem can be the failure of germinated seed to emerge from the soil, due to environmental stresses or to soil mechanical impedance. The final problem is mortality of emerged seedlings, due also to environmental stresses, plant competition, pathogen or pest attack.

fig. 4.2 *Urochloa* seed collected by seed harvesting ants

Table 4.3 shows the results of one very careful establishment study which defined the relative importance of these various factors in one specific problem area. The study was carried out with two small seeded grasses; *Chloris gayana* and *Panicum maximum*, whose seeds weigh about 0·2–0·3 and 0·5 mg respectively. The seeds were sown at 1·3 cm depth on black cracking-clay chernozem soil on the Darling Downs (lat. 28°S). This is a relatively dry cropping area where unreliable pasture establishment is one of the factors limiting the more extensive use of rain-grown sown pastures. These sowings were made on 17 occasions at intervals of two weeks throughout the spring, summer, and early autumn period. The results of these sowings may be divided simply into 12 sowings which germinated, and five which did not. Moisture stress was the main factor limiting field germination, and per cent germination depended on the depth to which soil was saturated with moisture by any rain after sowing. The seeds in the germinated sowings germinated nearly completely, (78 and 68 per cent for *C. gayana* and *P. maximum* respectively) and there were few seeds (3 to 4 per cent) left in the soil with any residual viability. On the other hand, seed in the non-germinated sowings retained quite good residual viability (83 and 47

Table 4.3 Average fate of seed after six weeks field exposure. (*Source:* Leslie, J. K. (1965). *Qld. J. Agric. Anim. Sci.* **22**:17.)

Parameter	Viable Seed (per cent)			
	Germinated sowings		***Non-germinated sowings***	
	Chloris gayana	*Panicum maximum*	*Chloris gayana*	*Panicum maximum*
Field germination	78	68		
Non-emergent seedlings	51	27		
Crust-impeded seedlings	2	1		
Seedling mortality	6	1		
Seedling survivors	7	24		
Residual viability	4	3	83	47
Seed not recovered		3	4	21
Non-viable seed	17	26	13	32
Number sowings (fortnightly, Sept.–April)	12		5	

per cent in *C. gayana* and *P. maximum* respectively). Seed spoilage was greater in *P. maximum.*

Table 4.3 shows that the main cause of establishment failure in the germinated sowings was not the development of soil crusts impeding the emergence of the seedlings, which had been thought to be a major cause of trouble. Only an average of 1 to 2 per cent of seedlings were affected in this way, although quite thick soil crusts formed if drying conditions were slow. The largest mortality class were 'non-emergent' seedlings. Close examination of the germinated seedlings showed that many of them did not penetrate the layer of dry soil crumbs below the surface crust. This was made worse by the penetration of light through the soil which favoured the appearance through the coleoptile of the primary leaf, which seemed to have less penetrating power. The smaller *C. gayana* seedlings (51 per cent) were more affected than *P. maximum* seedlings (27 per cent).

Mortality of emerged seedlings (1 to 6 per cent) was mainly due to heat and moisture stress. Mortality increased further if after some soil drying, a further rain and drying cycle occurred. Most of the seedlings emerged in the cracks between the hexagonal soil crusts; if a second series of hexagonal crusts formed these usually occurred in a new location, lifting the seedling and constricting it in the middle of its new 'island'.

The result of all the above was a 7 per cent establishment of *C. gayana* viable seed in sowings which germinated, and 24 per cent of *P. maximum* seed. This example has been discussed at length, as it illustrates well the special problems of establishing small seeded grasses on cracking soils. The principal difficulty lies in the dependence of establishment upon weather conditions after sowing. Small grass seeds cannot be safely sown deeper than about 1·3 cm on clay soils. By the time the surface of these soils has dried sufficiently to permit the passage of sowing machinery the seed must be placed in what has become a dry soil layer. Larger seeded grasses, e.g. *Sorghum almum*, or legumes such as *Glycine wightii* can if necessary be planted more deeply in a soil layer still moist from previous rains.

Environmental characteristics affecting establishment are listed below to provide a background against which agronomic establishment techniques may be discussed.

(a) Water relations

The effects of moisture availability on the initiation of germination and on the emergence, growth, and survival of seedlings have been mentioned. These processes are quite fundamental but they do not continually limit growth in all tropical environments. The basic conditions are the soil water energy, ease of transport to the seed or seedling, permeability of plant membranes, and rate of seed and soil

Table 4.4 Rate of mulch application and *Panicum maximum* establishment. (*Source:* Rickert, K. G. (1970). *Trop. Grasslds.* **4**:71.)

Rate of Mulch Application (kg/ha)	Density of *P. maximum* (plants/m²)	Yield of *P. maximum* (kg/ha)	Yield of Volunteer Species (kg/ha)
0	13	150	990
2 500	32	280	790
5 000	43	420	770
10 000	79	640	310
20 000	26	540	330

NB 1 lb per ac = 1·12 kg per ha

drying. Seed/soil contact also varies and can have a great influence. This is increased by rolling or by the removal of the bristles and glumes of some species, e.g. *Cenchrus ciliaris*. One of the effects of lime coating of seed is to increase the rate of water uptake from soil to seed, especially in surface broadcast seed. Pasture species differ in their capacity to germinate at low levels of water availability, but no good reports have been produced on this for tropical plants.

The rate of soil drying needs to be considered in relation to the rate of germination. In dry environments the optimum sowing time may be regarded as one where the widest gap between these two rates exists in favour of germination. Table 4.2 shows how rate of germination is dependent on temperature. Drying is similarly dependent and may be increased even more at high temperatures. Thus, even though rainfall incidence may be greatest in mid-summer, reliability of sowings may be better in the spring or late summer months when temperatures are cooler.

Surface cover, which is varied by stocking rate in established pastures and by the disposal of previous crop residues in ley farming systems, reduces the rate of soil drying. The retention or application of an organic mulch can favour pasture establishment through improved moisture relationships. Table 4.4 illustrates the results of some research work on the effects of different rates of organic mulch on the establishment of *Panicum maximum*. This was carried out by the Queensland Department of Primary Industries at the Brian Pastures Research Station, Gayndah (lat. 26°S). Seed at 4·5 kg per ha was raked into the surface soil and a chaffed pasture mulch placed on the surface. It will be seen that the density of sown grass increased from 13 to 79 plants per m² as mulch rate increased to 10 000 kg per ha. The density then decreased below the maximum when mulch was 20 000 kg per ha.

This effect was associated with better moisture availability and cooler soil temperatures. In three experiments the mean moisture content in the 0–1·2 cm soil layer in the first ten days after sowing averaged 8·1 per cent greater in 5 000 kg per ha mulch treatment compared with the control treatment. In the same periods the 09.00 hours temperature at 1·2 cm soil depth averaged 2·1°C cooler under the 5 000 kg per ha mulch treatment than under the control. Organic mulch has a high reflection coefficient for short wave radiation and low thermal conductivity. This reduces both soil temperatures and the amplitude of variation in soil temperature over 24 hours. On cloudy days mulch has little effect on temperature. On sunny days the rate of soil drying is probably reduced both by lower temperatures under the mulch, giving less heat energy available for moisture vaporisation, and by the greater resistance of the mulch to water vapour diffusion. The lesser effect of the highest mulch level was probably due to the greater physical barrier offered to seedling emergence. Table 4.4 also shows higher seedling dry matter yields of *P. maximum* in the mulched treatments, and decreased yields of volunteer species, mainly weeds.

There are two other aspects of moisture supply which have significant side-effects on establishment. The availability of some soil nutrients, particularly nitrogen, is increased by long severe drying periods before soil wetting occurs. This is known as the 'Birch effect' and is a particularly strong feature of soils with good content of organic matter. It seems to be related to the effect of drying on the structure of soil colloids. The bacterial population which grows on these after soil re-wetting is more effective in making nitrate available to the young seedling if the establishment rains occur after a long dry period. Consequently, unfertilised pasture sowings made at the beginning of the rainy season often have higher initial growth rates than sowings made towards the end of a rainy season, when the soil has been continually moist for long periods. A second side effect of rainfall is the leaching of water-soluble compounds in the seed coat or in the soil which inhibit the germination of some species. The occurrence of this effect is not very widespread.

(b) Gas exchange

A good supply of oxygen to the seed is necessary for germination. Low levels of oxygen availability also restrict seedling nutrient and water uptake. Atmospheric oxygen is normally sufficiently adequate for these purposes, except when moisture supply is excessive, due to poor drainage or to flooding. The availability of carbon dioxide in the atmosphere around the seed can help break dormancy of pasture species, especially in the presence of good oxygen supply. On the other hand, high accumulation of carbon dioxide within the tissues of the seed itself may form a barrier to germination of some dormant seeds.

(c) Temperature

Reference has been made to the positive effects of temperature on rate of seed germination and on soil drying. Some species such as *Chloris gayana* have a very wide temperature tolerance for germination. This enables successful sowings to be made in most seasons of the year. On the other hand it does not protect the seed from germinating at either very high or very low temperatures which may cause seedling mortality. Tropical pasture species have higher temperature optima for germination than temperate species, e.g. laboratory testing for viability is often 35°/20°C day/night temperature, compared to 30°/10°C or 20°C constant for rye grass. Work at the New South Wales Department of Agriculture Station at Wollongbar (lat. 29°S) has shown that germination of seed of *Glycine wightii* is reduced by soil temperatures commonly occurring on sunny days at that location. Exposure to temperatures of 37°C for 6 hours or of 45°C for 3 hours caused seed death during the early phases of the germination process, but was less damaging if high temperatures occurred after the radicle had actually emerged. Exposure to high temperatures on more than one day were particularly damaging, and reduced the rate of radicle extension of those seeds which did germinate.

Rhizobium may also fail to survive in hot soils, and cultivated sandy soils in the sub-tropics and tropics often record temperatures in excess of 45°C in their surface layers. These temperatures are known to be damaging to the successful inoculation of *Centrosema pubescens*. The timing of sowing in relation to weather, depth of sowing and provision of cover can all be varied to reduce this problem.

Alternating temperatures are favourable for the germination of most tropical grasses, especially if dormancy is present. *Stylosanthes humilis* responds in a similar way.

(d) Light

Germination of most grass seeds is favoured by light. The appearance of the common weed *Digitaria sanguinalis* after cultivation or after mowing a sward is associated with this effect. Only low levels of light intensity are needed to promote germination. The effect also depends on the reversible red/far-red phytochrome system. Red light promotes germination and far-red wave lengths inhibit germination. *Chloris gayana, Melinis minutiflora, Paspalum dilatatum* and *Setaria anceps* are examples of tropical grasses whose germination is increased by exposure to light.

(e) Soil physical characteristics

Earlier discussion mentioned the importance of soil/seed contact in promoting good water relations, and referred to two ways in which soil could reduce the emergence of seedlings sown at depth; the dry layer of crumbs beneath the surface or the formation of surface crust may prevent seedlings emerging. Sands, by virtue of their looser aggregation, offer less resistance to an emerging coleoptile. On the other hand some particular sands cement closely at the end of a wetting and drying cycle, offering a real barrier to small seedlings. The beating action of raindrops on bare soils may also form a soil skin resistant to further water entry and to seedling emergence. Clay soils with an exchange complex dominated by the sodium ion, e.g. solonised solonetz or solonchak soil types are well known for their poor physical characteristics and the failure of pasture establishment and growth which results. Displacement of the sodium ion with calcium by treatment with gypsum improves the crumb structure, but complete soil treatment is usually only economic under irrigation.

Establishment of surface sown seed requires that the radicle penetrates the soil surface downwards, and that the roots grow to sufficient depth early enough for the seedling to use deeper soil moisture reserves before the soil drying front makes these inaccessible. Pasture species differ greatly in their successful establishment from surface sowing; the legumes *Stylosanthes humilis, S. guyanensis,* and *Macroptilium atropurpureum* cv. Siratro reliably succeed where surface sown *Glycine wightii* reliably fails. Surface seeded *Melinis minutiflora* and *Panicum maximum* often form good stands where *P. coloratum* var. *makarikariensis* would fail. There has been no good research work which explains these differences in tropical species performance. The thicker radicles of legumes are generally less able to penetrate the soil surface than the finer grass roots, but if they have larger seeds with better energy reserves, they are not so quickly exhausted. Some restraint on the seed which holds it down in a vertical direction helps root penetration, and this accounts for the better success of seed broadcast on a rather irregular soil surface than on a perfectly smooth seed bed surface.

The rate of root extension is also influenced by soil physical characteristics, being faster in sandy soils, and also by plant characteristics. Research work at the University of Western Australia found that the rate of root penetration showed a range from 0·8 to 3·2 cm per day for 12 pasture species. The time to reach 20 cm depth was shorter for bigger seeds.

(f) Pathogens, pests and predators

Each locality will have its own range of diseases and pests affecting the seedling, and these problems often vary in their occurrence from season to season. Sowing time and planting procedures may often be changed to reduce the incidence of 'damping off' organisms such as *Pythium* spp. A dieldrin seed protectant slurry is necessary in southern Queensland to protect *Macroptilium atropurpureum* cv. Siratro from bean fly (*Melanagromyza phaseoli*), which can cause complete stand failure. Seed may need to be treated to repel or defeat predator seed harvesting ants, e.g. *Iridomyrmex* sp., in unreliable climates, where the period between sowing and germination may be lengthy.

(g) Competition

Pasture plant competition has been discussed in an earlier chapter, and brief reference is now made to some aspects related to seedlings. Usually the farmer tries to apply establishment techniques which will favour the sown species over unwanted weed species. On the other hand, he may want to create a grass-legume balance or a cash crop-pasture balance in a companion cropping situation. The main tools are seed size, rate and timing of sowing, fertiliser level and placement, previous land treatment, and grazing or cutting management.

Competition between newly emerged seedlings is rarely acute, but considerable early competitive stress may be placed on very young seedlings in sowings designed to upgrade established pastures. Perennial species with well-developed root systems are in a position to deny moisture and nutrients to young emerging seedlings. For example, the relative proportion of *Stylosanthes humilis* regenerating from seed in established pasture of *Cenchrus ciliaris* is positively related to the rainfall in that season, and in dry years *C. ciliaris* is strongly dominant. However, in a fully cultivated seed bed sown to both species, *S. humilis* shows better seedling establishment characteristics than *C. ciliaris* if moisture supply is critical. When upgrading existing pastures, either a sowing time should be chosen when the pastures in control are quiescent, or some check to their growth should be arranged.

The outcome of competition between seedlings is first influenced by the timing of germination, e.g. deep sown and late-emerging seedlings start at a disadvantage to fast-germinating weed seeds from the soil surface layer. It has already been mentioned that seed size is an important factor in deciding seedling shoot growth rate and root elongation rate. Once a leaf shades another leaf it reduces the photosynthetic rate of the latter and alters its capacity to assist root elongation, and nutrient and moisture uptake. A plant gaining a greater share of one environmental resource, e.g. light, is able to change other aspects of the environmental resources available to the dominated plant, and the effects added together. An intelligent understanding of these processes will result in management decisions leading to the competitive balance desired by the farmer.

The nature of the pasture-tree association determines whether it is desirable to remove trees in an open savanna woodland. Pastures may be established in contour strips sown between the trees, with some clearing to remove fallen logs and small bush regrowth. The success of this technique depends on the capacity of the sown species to colonise ground beyond the strips. Trees with well developed surface root systems, e.g. *Eucalyptus crebra* in Queensland, made it impossible to establish *Cenchrus ciliaris* close to the trunk, and it has been difficult to displace the natural *Heteropogon contortus* which grows there. On the other hand, in south-western Queensland, *C. ciliaris* has colonised the drip-ring area beneath *E. populnea* trees on low fertility soils, and has not grown well in the inter-tree areas. Table 4.5 shows that this is probably associated with better soil fertility conditions under the trees, which favour the more responsive *C. ciliaris*. The available phosphorus varied from 10 ppm, 10 m from the tree, to 160 ppm close to the tree trunk. Some transference of nutrients to stock camps under trees is expected. In this sample the improved fertility is probably associated with the capacity of trees to mobilise nutrients from surrounding soil and deeper layers and move them nearer the soil surface beneath the tree canopy.

Evidence from work of the CSIRO Division of Tropical Pastures at 'Kangaroo Hills' (lat. 19°S) in a dry environment indicates that tree removal can greatly increase the successful establishment and early stock productivity of pastures sown to *Stylosanthes humilis*. By contrast, scientists in the Northern Territory of Australia are impressed by the shading effect of a tree canopy maintaining leaf turgor of *S. humilis* for longer periods of the season than occur in a treeless situation. Most scientists favour tree removal (except for stock shelter areas, steep slopes and water courses), as this allows pasture productivity and stock numbers to be increased earlier in the establishment phase.

Table 4.5 Surface soil nutrient status under *Eucalyptus populnea*. (*Source:* Ebersohn, J. P. and Lucas, P. *Qld. J. Agric. Anim. Sci.* **22**:431.)

Distance from Tree Trunk (m)	"Available" P (ppm)	Exchangeable K (mol × 10^{-3} per 100 g soil)	pH
1·2	160	0·65	7·5
3·2	45	0·60	7·7
7·2	19	0·45	7·6
10·2	10	0·43	6·7

4 Agronomic techniques of pasture establishment

(a) Land preparation for sowing

Natural grasslands, or grasslands which are the result of fire or previous clearing, obviously have fewer problems of land preparation than forest lands. Removing trees is not always a good decision. Removal of trees from land too steeply sloping for the use of ground machinery has led to later pasture management difficulties, weed invasion, and erosion. Trees should not be removed from watercourses, except in carefully planned designs to preserve stable stream flow. One of the worst features of some land development programmes has been the piling of unwanted timber and debris in gullies and water-courses which has a bad effect on their flow characteristics. Well-managed pasture can provide good watershed run-off patterns and soil stability, but this is not possible on land subject to slips, which needs tree roots for anchorage.

The first step is the removal of timber which can be milled, using extraction techniques which do not concentrate run-off water and create erosion. In some countries the residual trees are then cleared by manual methods. The simplest technique is to cut over the area with knives or axes, burning the debris when dry enough, and hand-piling the unburnt debris for subsequent firing. In land to be cultivated the area is hand-grubbed to remove stumps or roots. Alternatively stumps and suckers which have started to grow again and regenerating seedlings are hand cut or treated with herbicide. Explosives may be needed to remove the larger tree stumps.

'Ring barking' trees is a technique where a continuous band of surface bark, perhaps 10 cm wide, is removed so that the upward movement of sap continues but the downward movement of assimilate to the roots is interrupted. This causes the slow death of the tree. A combination of manual and chemical methods is now more commonly practised. The use of a standard 13 cm blade axe with a cylinder delivering herbicide to the axe surface has been improved by narrowing the axe blade to 2·5 or 5 cm and giving it a straight chisel or concave edge. Many tree herbicide injectors use a long straight handle in contrast to the usual axe handle placed at right angles to the blade. The effective season of treatment, dosage rate, and application technique depend on local conditions. An illustration is given from research of the CSIRO Woodland Ecology Unit. *E. populnea* in southern Queensland may be controlled by injecting 2 ml of 1 per cent picloram as Tordon 50D or 105 (commercial preparations also containing 2,4,5-trichlorophenoxyacetic acid) at 13 cm centres around the circumference of the tree; height and season of application modify success very little.

Labour intensive methods are inappropriate in some countries, and trees are killed by mechanised (e.g aerial) herbicide application. Considerable success has been reported in treating tropical brushlands in Hawaii dominated by *Dicranopteris linearis* and *Melastoma malabathricum* with 4·4 kg per ha of the commerical compound Silvex; the application is repeated after eight months and the area subsequently burnt and seeded.

The most inexpensive system of mechanical removal uses two tractors travelling in parallel and linked with a chain which pulls over and partially uproots trees and shrubs. More uprooting occurs if the soil is moist. This method is preferred to tree-chopping machines which leave the basal trunk and roots intact. An additional shorter cable linking the two tractors may be used; this then rides forward, bending trees over before the chain strikes the base of the tree. Some people use a heavy rolling metal ball centrally attached to the chain, in order to prevent the chain from riding high up the tree trunks. In forests with a very variable size distribution the tractors operating in pairs will pass toward the centre to avoid very large trees, which may then be left as shade or removed by a single independent tractor unit equipped with a high tree-pushing ram and forward blade. If two tractors are unavailable a single tractor may operate in a circle with a chain anchored to a large tree.

In some pasture development operations it is unnecessary to remove residual tree roots. Sufficient tree fuel may be present to start a fire without moving the trees and pasture seed may be broadcast into the ash seed bed as soon as possible after firing. The fuel supply may be increased by seeding grasses such as *Melinis minutiflora* or *Chloris gayana*. Weeds and tree regeneration may be controlled later with ground misting equipment and destocking. Pushing the fallen trees with a bulldozer blade into windrows, root severance, exposure and raking are the most expensive parts in a clearing operation, and may be necessary if the area is to be fully ploughed early in its development. It is important to avoid 'dead' running one way; that is to say, the tractor should always be moving material when it is travelling. Either windrowing should be conducted as step-wise continuous travel in the same general direction, moving laterally one tractor width each time a blade-load is dropped, or reverse-travel time may be used to root-plough the area. The use of a blade with basal prongs will minimise carriage of surface soil to the windrow, if the operation is conducted skilfully. Contour placement of windrows is most desirable in order to reduce erosion during this vulnerable land preparation stage.

In wet heath lands some drainage works may be required. For flood irrigated pastures, bank layout, grading and construction are needed. If trees and stumps have been removed, the next phase is usually a deep ploughing. This is followed by successively shallower workings. The object is usually a fine, firm, even, level seed bed, in which the surface weed seeds have germinated and been destroyed. A most common fault is the production of a 'fluffy' seed bed. Rolling after sowing to consolidate the soil and provide better soil/seed contact does much to improve the reliability of establishment. It is tempting for farmers to use disc or roto-bladed instruments too much if there are large amounts of organic matter to be incorporated and these destroy soil aggregation more rapidly than tyned implements.

Experience suggests two alternatives. Pastures are often sown on sloping lands; vulnerability to erosion is reduced if a contour-worked rough seed bed is produced. Under dry conditions seed tends to establish in the furrow where moisture conditions are more favourable, and under very wet conditions plants establish better on the ridges. This type of seed bed is also more suitable for surface broadcasting, since it assists root penetration. The second alternative is minimal cultivation after the initial deep ploughing, and use of herbicides to kill weeds and weed seeds, avoiding the exposure of further weed seeds before sowing. The method which is actually used depends on the comparative costs and effectiveness of herbicidal and mechanical treatment in a specific farm situation. The practice of mixing herbicidal compounds such as trifluralin with soil three weeks before sowing has increased.

Table 4.6 Soil moisture regime and *Cenchrus ciliaris* emergence after seed treatment.

Soil Moisture Content (per cent) ***0–14 Days after Sowing***	**Soil Treatment** (per cent emergence)		
	Control	***Sulphuric Acid Treated***	***Hammer-milled***
5·0–1·4			
7·0–4·1	3	50	24
9·0–5·9	19	76	23
11·0–7·2	44	75	23

(b) Seed treatment

The purpose of seed treatment may be to break dormancy and improve seed flow characteristics, for rhizobial inoculation, or for seed protection..

Holding freshly harvested seed in storage will overcome dormancy problems in many grass seeds, e.g. *Cenchrus ciliaris* cv. Biloela reaches satisfactory germination levels after three months, but *C. ciliaris* cv. West Australian requires 12 months. Dry heat is sometimes used to improve germination in the laboratory, e.g. 10 days at 40°C is effective on new seed of *C. ciliaris.* Concentrated sulphuric acid is used on many seeds in the laboratory but this technique is too dangerous to seeds and man to recommend in field practice. A more acceptable alternative for legume seeds is treatment with hot water. Recommendations range from 80°C for two minutes for *Leucaena leucocephala*, to 75°C for six minutes for *Macroptilium atropurpureum* cv. Siratro, to 55°C for 20 minutes for *Stylosanthes guyanenis* and *S. humilis*. Seed should be dried and preferably sown immediately after treatment. Mechanical scarification designed to cause small cracks in legume seed coats is a further alternative. *Cenchrus ciliaris, Melinis minutiflora, Paspalum notatum* and *Stylosanthes humilis* may be hammer-milled to reduce dormancy or to improve seed flow. Under assured sowing conditions this improves the efficiency of establishment.

However, in one experiment in central Queensland 13 per cent of untreated *C. ciliaris* seeds produced plants, whereas only three per cent of treated seeds were successful. This arose from a single rain causing almost complete germination of treated seed. Later heat wave conditions caused complete seedling mortality and few viable seeds remained to produce plants when favourable climatic conditions for establishment occurred. Table 4.6 illustrates the effect of acid treatment or hammer-milling on the emergence of *C. ciliaris* under various moisture conditions. In this sandy soil, no seeds emerged under a soil drying regime starting at 5 per cent moisture and reaching 1·4 per cent moisture after 14 days incubation. Untreated seed did not display good emergence until moisture supply was very favourable (11–7·2 per cent), whereas treated seed exhibited maximum emergence at intermediate moisture availabilities.

The importance of an effective legume-*Rhizobium* symbiosis has been discussed in an earlier chapter, and the effect of mineral nutrition and management on the amount of atmospheric nitrogen fixed will be outlined later. Reference is made here to correct inoculation technique. Field sowings of many tropical legumes, e.g. *Macroptilium atropurpureum* cv. Siratro, *Pueraria phaseoloides, Stylosanthes humilis*, do not benefit from inoculation, since they nodulate promiscuously with all naturally occurring *Rhizobium.* Table 4.7 indicates the species, e.g. *Leucaena leucocephala, Lotononis bainesii*, for which inoculation with a correct, specific rhizobial strain is essential. Rhizobial cultures are available inexpensively in many countries, and peat cultures for all tropical pasture legumes which can be air freighted successfully around the world and which will store for several weeks are commercially available from countries such as Australia. Consequently lack of inoculum should not be a serious constraint on the success of establishment, provided suitable supply arrangements are made at the farmer level.

It is desirable to store inoculants in a cool place or under refrigeration; the key factor is to avoid exposure to sunlight. A peat inoculum may be simply mixed with water or milk, the seed thoroughly wetted with the mixture, and

Table 4.7 Inoculation requirements of tropical legumes. (*Source:* Norris, D. O. (1972). *Trop. Grasslds* **6**:59.)

Plant	Inoculation Requirement
Calopogonium mucunoides	Cowpea*
Centrosema pubescens	Specific
Desmodium intortum	*Desmodium*
D. uncinatum	*Desmodium*
Glycine wightii	Cowpea
Lablab purpureus	Cowpea
Leucaena leucocephala	Specific
Lotononis bainesii	Specific
Macroptilium atropurpureum cv. Siratro	Cowpea
Macroptilium lathyroides	Cowpea
Macrotyloma axillare	Cowpea
Medicago sativa	Specific
Pueraria phaseoloides	Cowpea
Stylosanthes guyanensis	Cowpea
S. guyanensis cv. Oxley	Specific
S. humilis	Cowpea
Trifolium repens	Clover
T. semi-pilosum	Specific
Vigna luteola	Cowpea

*Indicates a promiscuous species which will normally nodulate from native cowpea *Rhizobium* even if not inoculated

allowed to dry in the shade. A further refinement is to add a 'sticker', e.g. 1–2 per cent methyl cellulose (Methofas), Cellophas A, or 15 per cent gum arabic.

Making pellets of the inoculated seed with lime or rock phosphate has certain advantages. Seed and inoculant are separated from contact with fertiliser, water absorptive capacity is improved, bacteria are protected from sunlight, and more favourable nutritional conditions are provided. Tropical legume bacteria usually produce alkaline end-products. A lime pellet is specifically beneficial for *Leucaena leucocephala*, and for *Glycine wightii* in some circumstances, but a finely ground rock phosphate basis is usually best. On the other hand, temperate legumes such as *Medicago sativa* will establish much better if lime-pelleted, especially on acid soils. The following is a recommendation

fig. 4.3 Left-hand side: *Leucaena leucocephala* cv. Peru, inoculated with *Rhizobium:* right-hand side uninoculated native *Leucaena* on Bukit peninsula, Bali

for lime pelleting 50 kg *M. sativa*; similar methods are used for applying rock phosphate to other seeds. The ingredients needed are:

30–50 kg finely ground lime (plasterer's whiting)
2·5 kg gum arabic
0·5 kg peat inoculant for *M. sativa*
5 litre rain water

Gum arabic is dissolved in warmed rain water, and this solution is divided into two parts, to one of which peat inoculant is added. One third of the seed is placed in a small concrete mixer and one third of the gum arabic inoculant mixture is added to this and slowly agitated until the mixture is evenly sticky; some additional gum arabic may be necessary. The lime is divided into three parts and one part is added immediately to the mixture and agitated; a stirrer is used to break up lumps until all the seed is evenly coated. The seed is then spread to dry in the shade and the procedure repeated for fresh seed.

There are several other forms of pelleting for which increased seedling survival and growth are claimed. Grass seed may be pelleted to separate it from damaging fertiliser, for easier flow, or as a means of incorporating protectants. Protectants such as Lindane may be used to repel seed harvesting ants. Pelleting also prevents removal by ants.

(c) Sowing practice

Some factors influencing choice of sowing time such as the

influence of temperature on rates of soil drying and germination have already been discussed. One of the best recommendations made by an agricultural extension officer was to sow always 'just before three days of general rainfall'. Unless there are other reasons for delay, it is usually best to sow as early as possible in the rainy season, to obtain maximum growth and returns from the pasture in the year in which money is spent on establishment.

The use of sowing machinery which gives control of depth of sowing is desirable although many small farmers successfully establish pasture by hand broadcasting. The best depth for sowing depends on seed size. For the smallest seeded grasses, 1·3 cm is a desirable maximum depth, and shallower sowings are beneficial if control can be achieved. Work at the National Agricultural Research Station, Kitale, Kenya, indicated that for a wide range of pasture grasses, if seed was sown at a depth of more than 3 cm then there would be no emergence. *Stylosanthes guyanensis, S. humilis*, and *Lotononis bainesii* should be surface sown or no more than barely covered by the thinnest soil film feasible. Other legumes such as *Glycine wightii* will emerge satisfactorily from 2·5 cm depth; this gives better control of soil moisture about the seed. Many drills are built with a separate grass seed box attachment, so that seeds in a mixture can be sown at differing depths in the furrow; alternatively, seed may be mixed with fertiliser or sown through the fertiliser box at a depth adjusted differently from the seed delivery chute. Covering harrows, e.g. reversed diamond harrows, or a light chain, can provide controlled depth coverage. A press wheel which rolls directly above the sowing furrow, or post-sowing rolling give some depth control.

Recent ideas include the delivery of a narrow sand and/or gypsum treated band to provide a better physical environment for the germinating seedling. Heavy rates of herbicide may be employed if the seed is row-sown with activated charcoal placed in a narrow band in the furrow or on the soil surface above the seed. Bituminous mulches are effective in situations which are vulnerable to erosion, e.g. highway banks and drainage lines, where greater expense can be afforded.

(d) Rate of sowing

The importance of row spacing and control of plant density in giving maximum seed yields was mentioned in an earlier section. For normal pasture production, as distinct from seed production, there need be no fear that heavy seed rates will reduce output. In the first season, growth of the pasture, suppression of weeds, avoidance of erosion, the amount of nitrogen fixed by sown legumes, and the grazing returns are greater for high sowing rates, at least in moist environments. High sowing rates also greatly affect the production of later years. The alternative of low seeding rate is only successful if ground preparation and weed control are faultless, and if grazing is delayed. This permits stand thickening through natural seed drop and seedling regeneration, and through plant crown extension and runnering.

Table 4.8 Seeding rate *Lablab purpureus* production. (*Source:* Luck, P. E., personal communication.)

Seeding Rate (kg/ha)	Yield (kg/ha)	Additional Yield (kg DM/kg seed)	Break-even Seed Price ($/kg)*
5·6	1 720		
		250	3·00
11·2	3 100		
		116	1·39
22·4	4 400		
		31	0·37
44·8	5 100		

* Valuing yield at $12 per tonne

Figure 4.4a shows that the growth of an annual legume (*Stylosanthes humilis*) is density-dependent. Maximum seasonal yield varied from 1 100 kg per ha at the lowest density used to *ca* 8 000 kg per ha at the highest density. As indicated earlier, growth rate can be considered as the product of leaf area index (the capacity of the photosynthetic system) and the net assimilation rate (the efficiency of photosynthesis per unit of leaf surface present). Figure 4.4b indicates how the amount of leaf surface was the main control of growth, reaching a maximum of *ca* 5 m^2 of leaf above 1 m^2 of ground surface (= leaf area index of 5) at the highest density. On the other hand, because of greater shading within the canopy of high density swards, the net assimilation rate (Fig. 4.4c) was highest in low density swards, and fell away with time in all swards as growth proceeded.

A second example using the very large seeded *Lablab purpureus* (Table 4.8) shows clearly how increasing the seed rate shows a gradually reducing benefit. Seasonal

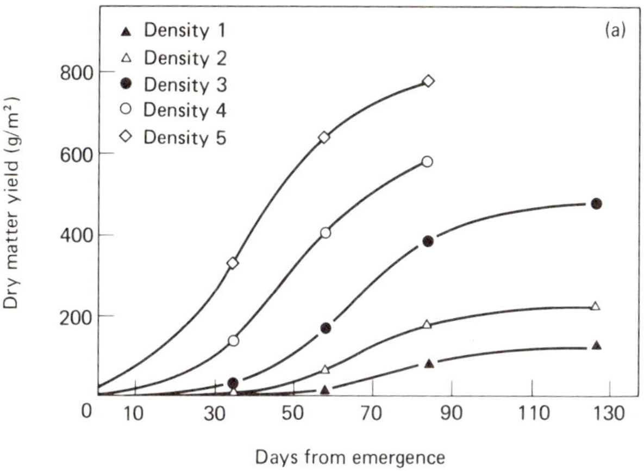

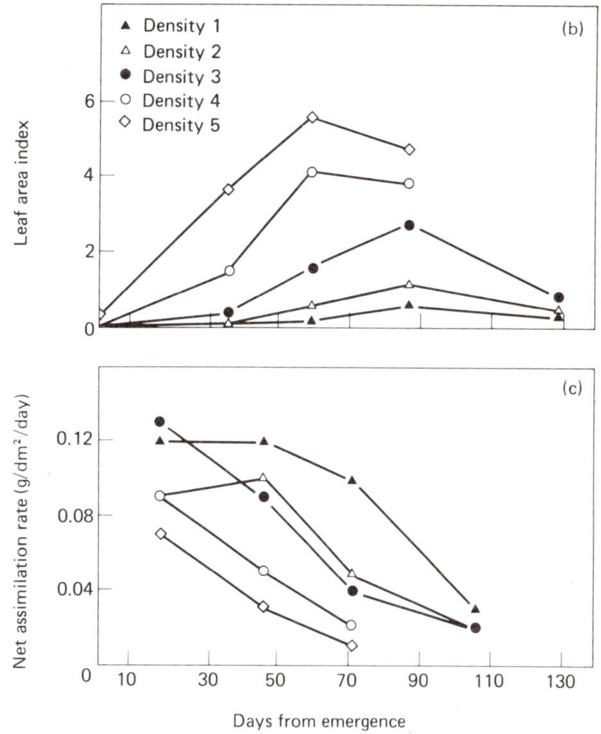

fig. 4.4 The effect of differing densities of *Stylosanthes humilis* (1:10, 2:50, 3:250, 4:1250, 5:6250, plants per m²) on (a) the growth of pasture (b) leaf area index, and (c) net assimilation rate. (*Source:* Rickert, K. G., and Humphreys, L. R. (1970). *Aust. J. Exp. Agric. Anim. Husb.* **10**:442)

fig. 4.5 A very sparse stand of the short lived *Sorghum almum* broadcast at 1 kg per ha, and of the perennial *Cenchrus ciliaris* (0·3 kg per ha) in its first year at Blackall, central Queensland

yield increased to the highest sowing rate tested, but the response to increasing sowing rate varied from 250 kg additional dry matter yield grown per kg additional seed sown in the 5·6–11·2 kg seed per ha range, to only 31 kg feed per kg seed in the 22·4–44·8 kg seed per ha range. If feed is valued at, for example, \$12 per tonne and no other cost factors are considered, a farmer would not lose money by increasing his seeding rate from 5·6 kg per ha if seed cost \$3 per kg. Above 23 kg seed per ha seed prices would need to be about 37¢ per kg or lower to break even on costs.

Seeding rates should be estimated on local cost and return factors. As a general rule it is suggested that the sowing rate of the total pasture mixture in tropical areas should never be less than 7 kg per ha, and preferably 10–12 kg per ha of good quality seed, in areas receiving more than 1 m annual rainfall; not less than 3 kg per ha in areas of 750 mm; and not less than 1 kg per ha in areas of 500 mm of rainfall. Choice of species has been discussed earlier; in mixed sowings on low fertility soils the legume sowing rate should be higher than the grass sowing rate.

Fig. 4.6 *Cenchrus ciliaris* assumes complete dominance in later years on cleared *Acacia cambagei* at Blackall, central Queensland

(e) Fertiliser placement and level

The competitive strength of the sown species compared to weeds is increased by placement of fertiliser in a band below and preferably to one side of the seed. Some fertilisers reduce pasture establishment if the seed is kept in contact. *Rhizobium* will be killed by acid single superphosphate, and a reduction in viability of *S. humilis* seed also occurs. Urea is damaging, especially if the biuret content is high. The fluoride content of superphosphate sometimes requires attention. These problems are greater on sandy than on clay soils.

The fertiliser needs of pastures and forage crops are discussed next in Part 5. Improved species are selected for their greater responsiveness to good nutrient supply, and productivity limitations are reduced by fertiliser use. Table 4.9 illustrates how the establishment of *Cenchrus ciliaris* on a poor sand depended on the use of a balanced fertiliser and/or cultivation. In natural uncultivated *Triodia* grassland, no establishment occurred, even with the use of fertiliser. With a minimum of land preparation (by disc harrowing), the use of fertiliser altered the competitive balance in favour of *C. ciliaris*. With deep ploughing which destroyed the native perennial grasses and released soil nutrients, *C. ciliaris* gave 70 per cent of the pasture yield, and complete dominance (94 per cent) was achieved if fertiliser was also applied.

(f) Vegetative propagation

Vegetative propagation is necessary for sterile plants, e.g. *Digitaria decumbens*, bred varieties of *Cynodon dactylon*, for plants with high cost or erratic seed supply, or for plants giving seed of low genetic stability, e.g. *Pennisetum purpureum*. The small farmer can use his own labour to develop planting material. Modern planting machinery makes low cost large area planting possible.

The first step is to obtain a supply of authentic material

Table 4.9 Cultivation and fertiliser effects on botanical composition of *Cenchrus ciliaris* sown areas. (*Source:* Edye, L. A. *et al* (1964) *Papers of Qld. Univ. Dept. Agric.* **1**(4):133.)

Treatment	**Botanical composition** (per cent by weight)				
	C. ciliaris	***Triodia pungens***	***Other Perennial Grasses***	***Annual Grasses***	***Herbs***
Uncultivated:					
–NPK	0	78	3	4	15
+NPK	1	63	10	23	3
Disc harrowed:					
–NPK	1	46	20	26	7
+NPK	30	27	22	20	1
Ploughed:					
–NPK	70	0	0	1	29
+NPK	94	0	2	2	2

and to grow a plant nursery area. Planting rooted sprigs is more reliable than using stem cuttings. To obtain sprigs, the area is mown and the top growth removed. Sprigs are then dug with a tyne or rotary cultivator, raked into windrows, collected, and transported to the development area. Furrows may be opened and the sprigs placed, roots covered and firmed in the furrow with some stem material above ground. This can be done by hand or by a variety of machines. The sprigs may be pressed into the soil with a disc coulter and culti-packer.

Stem cuttings are more dependent on favourable moisture conditions. Older, hardened stems are more reliable than young material. A well grown stand may be cut, in stages so that its height is gradually reduced, with a forage harvester, collecting stem pieces preferably bearing three nodes. The stems may be covered and made firm with planters designed for this purpose, or may be surface broadcast and covered with a disc drill seeding companion legumes.

Legumes may also be vegetatively propagated. Planting material from a stand of *Digitaria decumbens* containing *Lotononis bainesii* or *Desmodium heterophyllum* will give rise to a similar mixed stand.

Planting densities of about 0·7 m inter-row spacing and 0·3 m in the row give rapid first season production. Spacing on 2 m squares is satisfactory for runnering grasses if weed control is adequate and early season grazing of little importance.

5 Upgrading pastures with legumes

In many grazing situations low cost establishment of a well adapted legume in existing pasture is the management improvement plan which will give the greatest profit for a small investment. This arises where protein limits stock production and where soil nitrogen limits pasture growth. In the last chapter it was shown that *Stylosanthes humilis* is extremely useful in upgrading Northern Australian native pastures.

The success of the seeding operation usually depends on some (often minor) modification to the pasture environment, e.g. heavy grazing before seeding, which favours the sown legume. Annual grasslands in the Australian Northern Territory are vulnerable to firing immediately *after* the start of the wet season. Pastures are burnt when shooting has commenced, and *S. humilis* is successfully

fig. 4.7 *Stylosanthes humilis* seedlings in *Arundinaria* grassland, north-eastern Thailand

surface broadcast without cultivation after the fire. *S. humilis* is readily established in *Arundinaria pusilla* pastures in north-eastern Thailand. Throughout many parts of the tropical world *Imperata cylindrica* forms an induced disclimax which is cut for thatch and poorly used by animals. Although this grass forms a closed canopy the basal area actually occupied by the tussocks is low. Firing at the beginning of the rainy season exposes much bare ground which is easily colonised by *Macroptilium atropurpureum* cv. Siratro or *S. guyanensis* seedlings. This operation is more effective if mineral deficiencies (excluding nitrogen) are corrected. Growth of the companion legume improves stock acceptance of *Imperata* and may cause succession to other grasses.

Timing of seeding may be a critical factor. In subtropical Queensland, seeding of summer-growing perennial grass pastures with *Medicago sativa* is unsuccessful in the late summer but reliably successful after 25 mm rain in the late autumn or early winter when the companion grasses are quiescent. Removal of top growth, a single chisel ploughing, and drill sowing of inoculated seed with covering harrows behind is a recommended sequence of operations.

Minimal cultivation techniques are especially indicated on sloping lands where full seed bed preparation is both hazardous and difficult. The sod-seeder, in which seed is delivered within a shoe which cuts a narrow band in the turf, is sometimes successful with large seeded legumes. An alternative is to rotavate and seed narrow strips between undisturbed grassland; *M. atropurpureum* cv. Siratro is more successful than *Glycine wightii* in these partially prepared seed beds. The use of herbicides in lieu of mechanical cultivation has been canvassed for many years; preparations such as Dalapon applied at 6 kg per ha will check grass growth sufficiently to permit good *M. atropurpureum* cv. Siratro establishment. However, there are not many farmers in the tropical world who have accepted this technique and more economic techniques are usually available.

6 Sowing pastures with companion crops

Pastures may be sown after cash crops using the conventional sowing techniques described. A more profitable alternative is to sow pastures or forage crops with companion cash crops. This operation may have one or many of the following aims. The cost of pasture establishment is mainly absorbed in the cost of land preparation for crops. Many farmers in all parts of the tropical world are unwilling to clear land for pasture sowing, but will clear land for crops. Sowing pastures with crops suppresses weed growth in the inter-crop spaces, and affords protection against erosion. If used in an upland slash-and-burn situation the use of pasture legumes allows a fertility building phase which may be more efficient than in the jungle or 'bush-fallow' system. Forage legumes may be sown in order to achieve a short term improvement in dry season feed availability through the growth of additional feed and increased consumption of low protein cereal residues. On the other hand, the sown forages may form the basis of permanent pastures, or of a pasture phase of some years which alternates with a cash crop sequence. Finally, sowing a pasture legume in a cropping area where stock graze both cropping area and adjacent natural grassland creates an area from which a well adapted pasture legume will spread.

The aim is to devise a sowing technique which provides a minimum reduction in cash crop yield but still gives weed suppression and successful pasture establishment. This theme will be illustrated from work with *Stylosanthes guyanensis* cv. Endeavour (stylo) in upland crops, carried out as cooperative research between the University of Queensland, University of Khon Kaen, Applied Scientific

fig. 4.8 *Cynodon dactylon* established with a companion crop of *Avena sativa* at Muguga, Kenya

fig. 4.9 A pasture of *S. guyanensis* and *Melinis minutiflora* 16 months after broadcasting into upland *Oryza sativa* grown on the slash and burn system at Na Pheng, central Laos

Research Corporation of Thailand, and the Governments of Laos and Australia. The following variables may be manipulated.

(a) Pasture species

Pasture legumes are less likely to reduce companion cereal crop yields than pasture grasses. In one experiment in central Laos, upland rice sown in the ashes of land previously slashed and burnt suffered a 22 per cent reduction in grain yield if *Melinis minutiflora* and *Panicum maximum* were broadcast 34 days after rice sowing. On the other hand, sowing stylo caused no reduction of rice yield in these circumstances.

(b) Plant density

Upland rice has a faster seedling growth rate than stylo and the competitive balance favours the rice crop. High densities of stylo appreciably reduce rice grain yield, but moderate densities have little effect. Stylo sown 10 days after rice sowing to give a density of 25 plants per m^2, which can occur with 2 kg per ha of good seed, has little effect on rice grain yield. The rice crop reduces stylo yield. However, because tropical rice varieties tiller well at low density, the yield of stylo is not very sensitive to changes in rice plant density.

(c) Delayed sowing

The growth of a cash crop can be protected by delaying pasture sowing until the crop has gained an advantage. Delaying the sowing of stylo until the rice is established increases rice dominance. A delay in sowing stylo of only 10 days after rice sowing will prevent a reduction in rice yield.

(d) Crop maturity time

Stylo sown with an early flowering rice produces more yield than if sown with a late flowering rice variety which competes with stylo for a greater part of the growing season.

(e) Use of nitrogenous fertiliser

The application of nitrogenous fertiliser reduces the amount of stylo in the association. In one experiment the

fig. 4.10 *S. guyanensis* broadcast in upland *Oryza sativa* at Khon Kaen University, north-eastern Thailand. Right-hand side: mid-season crop variety; left-hand side: increased *S. guyanensis* growth with early crop variety

yield of stylo in the mixed association decreased from 3 400 kg per ha in the control treatment to 800 kg per ha where 80 kg elemental nitrogen per ha was applied. Figure 4.11 illustrates the even greater effects on the nitrogen yield of the association. Rice alone without nitrogenous fertiliser produced 34 kg per ha nitrogen in the tops; merely growing stylo with the rice crop increased the nitrogen yield to 97 kg per ha. This is a powerful and simple method of increasing the productivity of crop/pasture systems. On the sandy upland soils of north-eastern Thailand either the annual *S. humilis* or the perennial *S. guyanensis* may be sown in upland rice, sorghum, maize or kenaf.

(f) Plantation crops

The expansion of the livestock industry in plantation crop

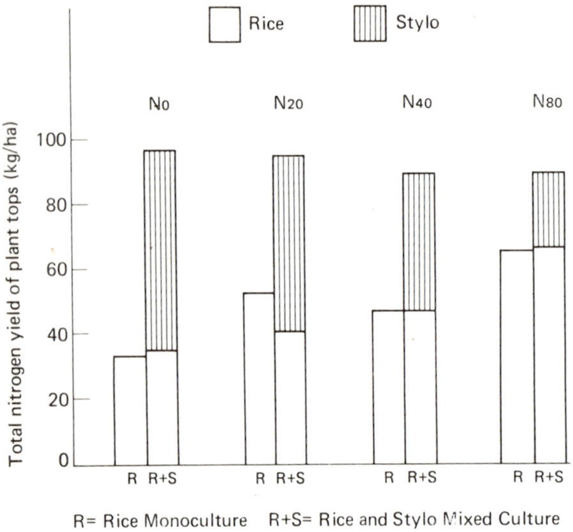

fig. 4.11 The total nitrogen yield in plant tops of rice monoculture and rice and stylo mixed culture at four levels of nitrogen application: 0, 20, 40, and 80 kg/ha. (*Source:* Shelton, H. M. and Humphreys, L. R. (1975). *Exp. agric.* **11**:103)

areas is one of the most promising developments in recent decades. Many plantation areas are introducing cattle which can be achieved at relatively low cost since the primary land development operation has already been carried out. This has led to greater income and to greater security of income. Control of weeds is aided, and the establishment of legume based pastures leads to increased soil fertility.

In young plantations, pastures may be first established in a central strip between the trees, so that competition is kept at a minimum. Oil palm is easily damaged by stock in the first two years, and after several years an almost closed canopy is formed which precludes much pasture growth. However, in the years in between it provides excellent grazing. Old coconut stands allow good pasture production, since light interception by the palms is slowly reduced as the palms age. Measurements in Bali (lat. 8°) at noon showed 77 to 80 per cent light transmission of mature palms 30 to 50 years old spaced 10 to 12 m apart. Percentage light transmission decreased if the sun's angle was lower, but these values indicate very good radiation conditions for pasture growth. Kapok plantations can also be used for grazing cattle. Fig. 4.14 shows pastures of *P. maximum* in a fruit orchard (lum-yai) near Chiangmai in northern

fig. 4.12 Bali cattle which had gained 0·5 kg per head per day during the dry season whilst grazing pastures of *Brachiaria decumbens, Centrosema pubescens, S. guyanensis*, and *M. atropurpureum.* Pastures were established by the University of Udayana under coconuts at Sangian, west Bali

fig. 4.13 Pastures of *Setaria anceps* and *M. atropurpureum* established in young oil palm at Serdang, west Malaysia

Thailand. These pastures are managed on a cut-and-carry system, and and are fed to housed dairy stock; this enterprise has proved very profitable.

The disadvantages of companion cropping arise if physical damage to trees occurs, or if the plantation crop yields are decreased due to pasture competition for water or nutrients. Competition for moisture is not a significant factor in the truly humid tropics. Where moisture stress is occasionally experienced, one needs to consider whether clean cultivation is either practicable or profitable. Sown pastures are a better alternative to weeds, and trees will at least benefit from the nitrogen added to the system. Sod forming grasses, e.g. *Brachiaria decumbens*, are preferable to tall bunch grasses such as *P. maximum* since the coconuts are more visible on the ground. Animals may cause damage to soil structure if they are grazed on heavy soils under very wet conditions. Soils are more easily damaged during the early phases of pasture establishment before a close sod is formed. Confining animals to upland situations or to pastures on light soils during periods of heavy rainfall or waterlogging reduces this problem.

Added inputs of non-nitrogenous fertiliser are needed to make sure that competition for nutrients is kept as low as possible. The nutrients retained by meat-producing animals grazed in the plantation are a minor loss. However more substantial nutrient transfers occur with milk-producing animals. A cut-and-remove feeding system causes a greater nutrient transfer but will reduce animal damage to both soil and trees.

fig. 4.14 *Panicum maximum* grown under a cut-and-carry system for dairy production in a lum-yai orchard near Chiangmai, northern Thailand

7 Grazing management of the growing pasture

Grazing management is discussed in the final chapter of this book. At this point it is only desired to refer to the two extremes of grazing pressure needed in different establishment situations.

Quite high early grazing pressure is desirable if low-growing, shade-intolerant legumes, e.g. *S. humilis, Lotononis bainesii, Desmodium heterophyllum, Trifolium repens,* or *T. semi-pilosum*, are sown. This is to prevent companion grasses from overtopping the legumes and restricting their growth. *S. humilis* is not usually selectively eaten when green, and high stock numbers are needed from sowing time to ensure its successful growth in competition with tall bunch grasses.

The other extreme is represented by the situation where vine-type climbing legumes, e.g. *M. atropurpureum* cv. Siratro or *Glycine wightii* are established at low density and many tall annual weeds develop. One plan is to exclude stock until the climbing legumes have smothered and shaded the weeds, seeded down, and grown a strong crown and root system. Intermediate situations may be recognised where careful grazing during the first season encourages grasses to stool out and discourages weeds which are edible or easily damaged by trampling. Careful grazing commences nutrient recycling, and provides some first year product return from the pasture.

No simple recipes are available for tropical pasture establishment. The intelligent farmer who understands the nature of the processes at work will make decisions which give him maximum long term gains from the pasture.

Further reading

Asher, C. J., and **Ozanne, P. G.** (1966) Root growth in seedlings of annual pasture species. *Plant and Soil, 24*:423.

Cameron, D. F. (1967) Hardseededness and seed dormancy of Townsville lucerne *(Stylosanthes humilis)* selections. *Aust. J. Exp. Agric. Anim. Husb.*, 7:237.

Ebersohn, J. P., and **Lucas, P.** (1965) Trees and soil nutrients in south-western Queensland. *Qld. J. Agric. Sci., 22*:431

Edye, L. A., Humphreys, L. R., Henzell, E. F., and **Teakle, L. J. H.** (1964) Pasture investigations in the Yalleroi district of Central Queensland. *Papers of the Queensland University Department of Agriculture, (4)*:133.

Harty, R. L. (1972) Germination requirements and dormancy effects in seed of *Urochloa mosambicensis. Trop. Grasslds., 6*:17.

Leslie, J. K. (1965) Factors responsible for failures in the establishment of summer grasses on the black earths of the Darling Downs, Queensland. *Qld. J. Agric. Sci., 22*:17.

Leslie, J. K. (1968) Seed and seeding physiology. *Proc. Aust. Grasslds. Conf.* CSIRO Melbourne *2*:57.

Luck, P. E. (1972) In search of cheaper pastures. *Trop. Grasslds., 6*:77.

Ludlow, M. M., and **Wilson, G. L.** (1972) Relationship between seed and seedling dry weight of tropical pasture grasses and legumes. *J. Aust. Inst. Agric. Sci., 38*:65.

McWilliam, J. E., and **Dowling, P. M.** (1970) Factors influencing the germination and establishment of pasture seed on the soil surface. *Proc. XI Internat. Grassld. Congr.* Surfers Paradise Australia p. 578.

Murtagh, G. J. (1970) Effect of temperature on the germination of *Glycine javanica. Proc. XI, Internat. Grassld. Congr.* Surfers Paradise, Australia, p. 574.

Norris, D. O. (1972) Leguminous plants in tropical pastures. *Trop. Grasslds., 6*:159.

Prodonoff, E. T. (1967) *Seed Testing in Queensland.* Queensland Department of Primary Industries, Brisbane, pp. 58.

Rickert, K. G. (1970) Some influences of straw mulch, nitrogen, fertiliser and oat companion crops on establishment of Sabi panic. *Trop. Grasslds., 4*:71.

Rickert, K. G., and **Humphreys, L. R.** (1970) Effects of variation in density and phosphate application on growth and composition of Townsville stylo *(Stylosanthes humilis). Aust. J. Exp. Agric. Anim. Husb., 10*:442.

Roberts, E. H. (1972) *Viability of Seeds.* Syracuse University Press, p. 448.

Robertson, J. A., and **Moore, R. M.** (1972) Thinning *Eucalyptus populnea* woodlands by injecting trees with chemicals. *Trop. Grasslds., 6*:141.

Shelton, H. M., and **Humphreys, L. R.** (1972) Pasture establishment in upland rice crops at Na Pheng, Central Laos. *Trop. Grasslds., 6*:223.

Stocker, G. C., and **Sturtz, J. D.** (1966) The use of fire to establish Townsville lucerne in the Northern Territory. *Aust. J. Exp. Agric. Anim. Husb., 6*:277.

Part 5 Soil fertility and fertiliser needs of tropical pastures

Pastures are usually grown on lower fertility soils which are not well suited to cropping. Hence soil fertility problems are found in most tropical pasture situations. The use of fertilisers is a good way of increasing pasture growth and nutritive value, and thus for increasing animal numbers and individual animal performance. Fertilisers are often needed to increase the efficiency of the legume–rhizobial symbiosis or association on which the increase of nitrogen in many pasture systems depends. For districts where fertiliser supply is uncertain or fertiliser prices are high, pasture species need to be identified which are efficient in extracting nutrients from low fertility soils or which grow better than other plants with low concentrations of nutrients in their tissues. Cash cropping systems may depend upon the use of pastures in the crop rotation to maintain or add to soil fertility.

In this chapter the role of the various plant nutrients in pasture production is mentioned, and the pasture response to added nutrients is outlined. The diagnosis of mineral deficiencies and toxicities, and the principles of pasture fertiliser practice are discussed. The effect of nutrient supply on botanical composition and the influence of grazing on nutrient cycling in the soil–plant–animal system are considered. Finally, factors affecting the legume–rhizobial symbiosis and the success of crop–pasture rotations are discussed.

1 Plant nutrients and their role

Carbon, hydrogen and oxygen (from carbon dioxide and water) form over 90 per cent of the dry matter content of tropical pasture plants. The main mineral constituents which affect farm fertiliser policy are nitrogen (N), phosphorus (P), potassium (K), sulphur (S), calcium (Ca), and magnesium (Mg). The principal micronutrients are copper (Cu), zinc (Zn), manganese (Mn), iron (Fe), boron (B), and molybdenum (Mo). An element is essential if a plant cannot complete its life-cycle without it, and if it forms part of the molecule of an essential plant constituent. For instance magnesium is a part of the chlorophyll molecule on which photosynthesis depends. All the above elements are essential; in addition chlorine (Cl) is essential, but is sufficiently abundant in the atmosphere to prevent field occurrence of chlorine deficiency in pastures.

Some other elements are essential for particular plants, or influence plant growth. Cobalt (Co) is essential for the legume–rhizobial symbiosis, and affects the synthesis of the vitamin B_{12}, but field responses of legume growth to cobalt are rare. Sodium (Na) is needed for some plants living in saline situations; it may also ameliorate potassium deficiency. Plants respond to silicon (Si) on some tropical soils, and this is probably due to silicon reducing the toxicity of other minerals.

A full description of the role of essential nutrients would need a biochemistry text-book, since all aspects of plant growth and differentiation processes are involved. Nitrogen is a constituent of all proteins, and a high content of non-protein nitrogen in plants is usually an indication of some mineral imbalance, (e.g. sulphur deficiency), which affects protein synthesis. Nitrogen is part of amino acids and amides, the nucleic acids which play a vital part in cell metabolism, and of coenzymes which control the rate of chemical reactions in the plant. Phosphorus plays a key role in the energy transfers of all living cells, and regulates many enzyme processes. Potassium also activates many enzyme systems in the plant. Calcium is the main cation present in the centre of the cell wall, and is a key factor in the structural organisation and strength of the plant; it is also involved in some enzyme reactions. Magnesium is involved in numerous enzyme reactions, especially energy reactions involving phosphate.

Although the micronutrients form a very small fraction of plant tissues, their role in growth and development is just as important as the macronutrients. Deficiences cause interruptions to growth just as damaging as those caused by a shortage of major nutrients. All the micronutrients are involved in enzyme systems of one kind or another as activators, cofactor or regulators. Boron regulates carbohydrate transport in the plant. Molybdenum is essential for nitrate reduction before protein synthesis, and the accumulation of nitrate in the plant is one symptom of molybdenum deficiency. Molybdenum is also required for nitrogen fixation by *Rhizobium*. Manganese is involved in energy transfers in the plant. Iron is needed for chlorophyll synthesis, and is a component of cytochromes, which influence energy relations. Zinc deficiency has a great effect on the growth substance auxin, which controls inter-node growth. Copper is involved in protein synthesis, and is also a

fig. 5.1 *Stylosanthes guyanensis* at Gadjah Mada University, Jogjakarta Indonesia. Left-hand side, no added potassium: right-hand side, with potassium chloride

cytochrome constituent.

The more common deficiency symptoms which are characteristic of these elements are outlined later in the chapter.

2 Response to added nutrients

(a) Limiting nutrients and their interactions

Sometimes the growth of a tropical pasture is limited by the supply of a single nutrient, e.g. potassium, and in this situation the pasture manager simply needs to find out the amount of potassium required to remove this limitation, and to decide how much potassium fertiliser gives the best return for the money spent. A more usual situation is to discover that more than one nutrient is limiting pasture growth, e.g. both phosphorus and potassium may be needed. Some very infertile soils may have several deficiences, e.g. legume based pastures in the Queensland coastal lowlands need added P, K, S, Ca, Cu, Zn and Mo. These situations require a balanced approach to fertiliser needs, in which all deficient nutrients and their interactions with one another are considered.

We must also recognise that in cases where a single nutrient deficiency is the main problem, the level of this nutrient which gives maximum pasture growth depends upon the level of availability of other nutrients. If these are in good supply, a higher level of the deficient nutrient is needed. As one nutrient deficiency is met, so the growth of pasture becomes limited by the second most important nutrient deficiency, and so on.

Table 5.1 Superphosphate and potassium chloride effects on pasture yield. (*Source:* Jones, R. J. (1966). *Proc. Trop. Grassld. Soc. Aust.* **6**:23.)

Level of Superphosphate (kg/ha)	**Relative Yield *** *Level of Potassium Chloride* (kg/ha)		
	0	190	380
0	100	112	113
190	109	124	132
380	111	133	133

* 100 is for pasture without fertiliser

When two nutrients are added there may be what is called a 'positive interaction' or in other cases there may be a 'negative interaction'. A positive interaction occurs where the response to the application of both nutrients is greater than the sum of the responses when they are separately applied. This is illustrated in Table 5.1, which shows data from an experiment on a podzolic soil near Gympie, Queensland (lat. 26°S, 1 140 mm). A pasture of *Setaria anceps, M. atropurpureum*, and *M. lathyroides* received three rates of single superphosphate, and three rates of potassium chloride application, in all combinations. With a yield of 100 for pasture without fertiliser, the relative yields of pasture with the various fertilisers were calculated. In the absence of applied superphosphate, the yield of pasture receiving 380 kg per ha potassium chloride was 113. In the absence of applied potassium chloride, the yield of pasture receiving 380 kg superphosphate per ha was 111. However, if both fertilisers were applied at these levels the yield was 133. This indicates a positive interaction between the two fertilisers, which for the 380 kg per ha levels is calculated as

$$\begin{aligned}[133 - 100] - [(113 - 100) + (111 - 100)] &= 33 - [13 + 11] \\ &= +9\end{aligned}$$

This effect of one nutrient reinforcing the effects of another is a great economic benefit since it leads to increased growth per unit of fertiliser applied. Interactions of phosphorus with potassium, nitrogen, sulphur or molybdenum occur quite commonly on tropical pastures. On very phosphorus deficient soils the application of nitrogen fertiliser without adding phosphorus may actually decrease pasture growth, and this is clearly a very wasteful practice. Positive interactions of phosphorus with sulphur or phosphorus with molybdenum are common in pasture legumes, and correcting these deficiences leads to increased nitrogen fixation and to a higher plant protein content. Positive nitrogen with sulphur interactions in pasture grasses also indicate the important role of sulphur in nitrogen metabolism.

Table 5.2 Effect of calcium carbonate and molybdenum on yield of *Macroptilium lathyroides*. (*Source:* Teitzel, J. K., and Bruce, R. C. (1972). *Aust. J. Exp. Agric. Anim. Husb.* **12**:49.)

	Yield (grams per pot)	
	Without Calcium Carbonate	***With Calcium Carbonate***
Without molybdenum	3·2	5·3
With molybdenum	4·3	5·5

A negative interaction between nutrients is where the response to the application of both nutrients is less than the sum of the responses when they are separately applied. This occurs when two nutrients have similar functions and may substitute for one another to some extent. Interactions of calcium and magnesium are an example since these are often negative. A similar effect occurs where the application of fertiliser alters the availability of another nutrient to the plant. A classic example is the application of lime, which by altering soil pH, increases the availability of molybdenum. Table 5.2 shows the response of the legume *Macroptilium lathyroides* to added calcium carbonate (limestone) and to molybdenum on a basalt-derived krasnozem soil in coastal north Queensland. Yield was increased from 3·2 to 4·3 g by adding molybdenum, and to 5·3 g by adding calcium carbonate. However, if both molybdenum and calcium carbonate were added the yield was 5·5 g, nearly the same as without molybdenum. The interaction of the two nutrients is calculated as

$$
\begin{aligned}
|5{\cdot}5 - 3{\cdot}2| - |(5{\cdot}3 - 3{\cdot}2) + (4{\cdot}3 - 3{\cdot}2)| &= 2{\cdot}3 - (2{\cdot}1 + 1{\cdot}1) \\
&= -\,0{\cdot}8
\end{aligned}
$$

and is clearly a negative interaction.

The calcium content of the plant tissues in all treatments was well above 1 per cent, indicating that calcium was not needed as a nutrient. The nitrogen content of the plant was increased by both calcium carbonate and by molybdenum, indicating that either treatment increased the efficiency of the rhizobial symbiosis. The lack of response to molybdenum in the presence of added calcium carbonate was probably due to calcium carbonate making molybdenum more available in the soil. The better yield of the plus calcium carbonate treatment (5·3 g) relative to the plus molybdenum treatment (4·3 g) must have been due to some other effect of calcium carbonate. It may have reduced the available manganese level which was otherwise too great and toxic.

The use of lime on tropical pastures is often unnecessary. One classic case occurred in northern New South Wales where farmers were applying 2 tonnes per ha of lime to pastures. It was then found that 200 kg per ha of lime drilled close to the pasture legume seed being sown gave the same effect. Later studies showed that the application of 60 g per ha of molybdenum gave an identical effect, since the response to lime was due to better molybdenum availability. It is important to distinguish between tropical pastures and temperate pastures when deciding if lime is to be used since many tropical pasture species and tropical soils respond differently. The application of lime in many tropical situations has proved costly and disastrous, and has in some circumstances caused substantial decreases in pasture yield.

As indicated earlier, calcium may function as a nutrient, controlling cell differentiation and the unity of plant membranes. Tropical pasture legumes, which have evolved on highly leached soils generally of low mineral level, have been shown by CSIRO scientists to have a superior capacity to extract calcium from low calcium soils. One experiment showed that four temperate legumes grown on a low humid gley soil without added calcium gave yields which ranged from 0·2 to 6·5 per cent of their maximum yields. Under the same conditions, five tropical pasture legumes gave yields which ranged from 25 to 68 per cent of maximum yields. *L. bainesii* is able to grow at very low levels of calcium, and a calcium content in the shoots of only 0·2 per cent is apparently sufficient for normal growth.

The addition of lime may also improve the physical structure of the soil, improving its aeration and water-holding characteristics. Lime addition will increase soil pH, which will affect nutrient availability. As indicated above it will increase the availability of molybdenum and reduce toxic levels of manganese and aluminium (Al). Phosphorus availability is also usually improved, especially on high aluminium soils. On the other hand, it may induce deficiencies of zinc, manganese, copper, boron and iron. This is quite common on alkaline or neutral soils.

Table 5.3 Lime application and legume/grass balance. (*Source:* Olsen, F. J., and Moe, P. G. (1971). *E. Afr. Agric. For. J.* **37**:29.)

Sown Legume	**Per cent Legume** *Lime Application Rate* (tonnes/ha) 0	2·5	5	10
Medicago sativa	32	49	47	59
Desmodium intortum	80	88	85	78
Stylosanthes guyanensis	76	72	76	82

The differential response of temperate and tropical pasture legumes to addition of lime is illustrated in Table 5.3. On this latosol with pH 5·6 at Kabanyolo, Uganda, the content of *Medicago sativa* in the sward after two years of treatment was 32 per cent in the untreated plots, and 59 per cent where heavy applications of lime had been made. On the other hand, dominance of *Desmodium intortum* and of *Stylosanthes guyanensis* were maintained on the untreated plots, and the degree of dominance was not affected by lime level.

(b) Yield responses

Figure 5.2 indicates the usual course of pasture yield response to increasing level of nutrient supply. Over the

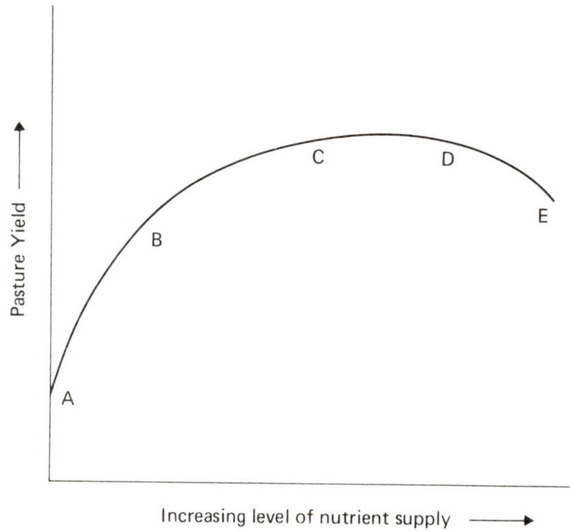

fig. 5.2 Diagrammatic indication of yield response to added nutrient

range AB where nutrient deficiency is serious, the yield response to added nutrient is high and linear. As a sufficient level is reached over the range BC, the law of diminishing returns operates, and the yield response to added nutrients is curvilinear and it is less for each step in which nutrient is added. Over the 'luxury' uptake range CD, yield forms a plateau and is unaffected by nutrient supply. Finally, additions of nutrient are toxic over the range DE and reduce pasture yield to levels less than the maximum.

This generalised picture of yield response is not correct for all situations. In some cases the region AB may show diminshing response per unit of fertiliser at very low levels of fertiliser application. For instance, in P-fixing soils, the first stage of fertiliser addition may interact with the soil and little may be available for plant growth.

The concept of diminishing returns was mentioned earlier in Part 3. Table 3.17 showed that the yield response of *Pennisetum americanum* was 30 kg dry matter per kg N at the 100 kg N per ha level, decreasing to 17 kg per kg at the 600 kg N level.

The toxic effect of nutrients is indicated in Figure 5.3. As mentioned earlier, calcium is a nutrient required only in minute amounts; at quite low levels it can cause decreased growth. Potassium is often supplied to pastures as potassium chloride (muriate of potash). Figure 5.3 compares the effect of two sources of potassium, potassium sul-

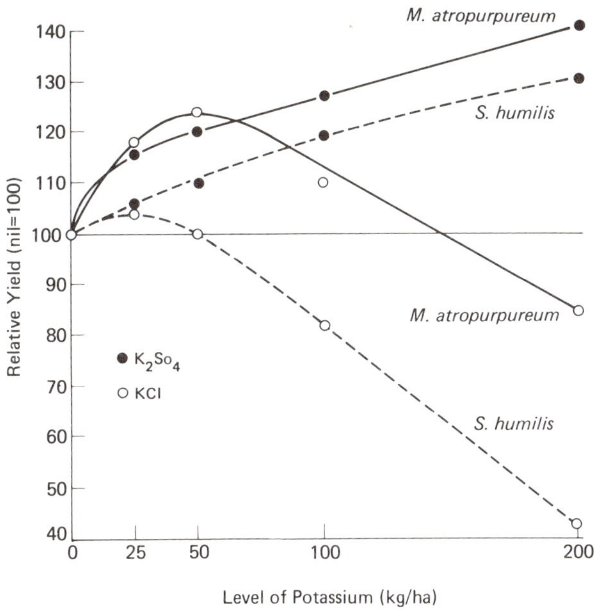

fig. 5.3 The relative yield of *Macroptilium atropurpureum* and *Stylosanthes humilis* with increasing levels of potassium applied as potassium chloride or potassium sulphate to a podzolic soil. (*Source:* Hall, R. L. (1971). *J. Aust. Inst. Agric. Sci.* **37**:249)

fig. 5.4 *Chloris gayana* at Kitale, Kenya. Foreground: no added fertiliser; back left-hand side: with added phosphates; centre left: depressed growth with added nitrogen but no phosphorus; right-hand side: increased growth with balanced addition of nitrogen and phosphorus

phate and potassium chloride; in this experiment sulphur application was equal in all treatments. It will be seen that yield of *Macroptilium atropurpureum* and of *Stylosanthes humilis* increased up to the highest level of potassium if this were applied as potassium sulphate. However, if more than 50 kg potassium per ha was applied in the form of potassium chloride to this podzolic soil, drastic reduction in yield occurred, especially with *S. humilis*.

The responses of tropical and temperate grasses to varying levels of nitrogen and phosphorus have been compared in many studies. Generally, tropical grasses yield better than temperate grasses under conditions of low nutrient supply due to both their superior capacity to extract nutrients and to their lower nutrient concentration. At comparable yields this gives a yield response advantage to the temperate grasses, but since yield responsiveness is closely related to growth potential, the superior yields of tropical grasses in the field lead to greater dry matter returns from added nutrients. The responsiveness of temperate pasture grasses to added nitrogen is usually in the range of 10 to 30 kg additional dry matter yield per kg of nitrogen applied. The tropical grass response is more usually in the range 20 to 50 kg dry matter per kg of N.

(c) Effect of fertilisers on nutritive value

Pastures are usually grown to be eaten, and the response to fertilisers must be assessed in terms of both growth and nutritive value. For some elements, animals' needs are not related to plant needs. For instance; the content of sodium and of cobalt may limit animal performance but not affect pasture growth. In many situations fertiliser application will increase the content of minerals or of compounds of high biological value which increase animal output. Table 5.4 indicates the effect of 260 kg per ha of single superphosphate on the mineral content of a natural pasture oversown with *Stylosanthes guyanensis* at Serere, Uganda (lat. 2°N, 1 100 m in altitude, 1 380 mm rainfall). Single superphosphate contains P, S and Ca, and it will be seen that fertiliser application increased the P and S content of

Table 5.4 Superphosphate effect on nutrient content (%) of pastures. (*Source:* Stobbs, T.H. (1970). *E. Afr. Agric. For. J.* **35**:128.)

Species	Nutrient	Without Superphosphate	With Superphosphate
Stylosanthes guyanensis	Nitrogen	1·67	2·19
	Phosphorus	0·16	0·20
	Sulphur	0·07	0·13
Imperata cylindrica	Nitrogen	0·73	0·86
	Phosphorus	0·12	0·19
	Sulphur	0·05	0·10
Other grasses	Nitrogen	0·83	1·08
	Phosphorus	0·10	0·15
	Sulphur	0·06	0·13

Table 5.5 Superphosphate rate and *Stylosanthes humilis* response. (*Source:* Shaw, N. H., Gates, C. T. and Wilson, J. R. (1966). *Aust. J. Exp. Agric. Anim. Husb.* **6**:150.)

Superphosphate Application Over Two Years (kg/ha)	Dry Matter Yield (kg/ha)	Chemical Composition	
		Nitrogen (per cent)	*Phosphorus* (per cent)
0	2 740	2·36	0·06
190	4 850	2·73	0·09
380	5 480	3·20	0·16
755	6 500	3·28	0·21

Table 5.6 Fertiliser nitrogen and seasonal content of *Hyparrhenia rufa*. (*Source:* Tergas, L. E., Blue, W. G., and Moore, J. E. (1971). *Trop. Agric.* **48**:1.)

Date	Level of Nitrogen Applied (kg/ha)		
	0	**75**	**150**
	Per cent Nitrogen		
November 30	0·80	1·00	1·44
January 5	0·44	0·37	0·47
February 9	0·27	0·24	0·28
March 5	0·21	0·22	0·30

legume and grass species to levels which were safe for animal production. The N content of *S. guyanensis* was increased from 1·67 to 2·19 per cent, a great increase in protein value. Similarly, Table 5.5 shows the positive effect of superphosphate on *S. humilis* N and P content. The increase in N content from 2·36 to 3·28 is the result of a great increase in symbiotic N-fixation efficiency and protein synthesis. Nitrogen content has been raised to levels similar to those of legumes with previously superior reputation for nutritive value.

The effect of nitrogen fertiliser on the nutritive value of pastures is more difficult to decide. The application of nitrogen fertiliser usually increases grass nitrogen content, especially at high levels of nitrogen application, say above 150 kg nitrogen per ha. However, at lower levels of fertiliser application the rapid growth caused by nitrogen fertiliser dilutes the nitrogen content rapidly unless further applications are made. In the field this effect is often made worse by a difference in grazing pressure. Unless stock numbers are increased to consume the extra pasture grown as it is grown, the nitrogen fertilised grass will reach a more advanced maturity stage more quickly and nitrogen content may even be reduced when compared with an unfertilised pasture on the same date. Nitrogen fertiliser will also not necessarily increase the nitrogen content of grass in the dry season.

The data in Table 5.6 are drawn from an experiment in Guanacoste Province in the wet-dry Pacific region of Costa Rica. *Hyparrhenia rufa* pastures were cut on October 13, and nitrogen fertiliser applied at three rates. Rain ceased on December 10 and the nutritive value of the standing pasture was assessed during the dry season. Cellulose digestibility decreased with time. At the end of the wet season, there was an initial response in nitrogen content from added nitrogen fertiliser of 0·8 per cent nitrogen in the control plots relative to 1·44 per cent nitrogen where 150 kg N/ha was applied. However, treatment differences in the standing pasture decreased subsequently and nitrogen content (0·2 to 0·3 per cent) became well below animal maintenance needs.

Under conditions of serious nitrogen deficiency, nitrogen fertiliser can increase the nutritive value of forage by correcting protein deficiency. Table 5.7 shows the effect of applying an additional 57 kg nitrogen per ha as urea (high nitrogen treatment) to a sward of *Digitaria decumbens* 28 days before cutting hay. The experiment was conducted on low humic gley soil at Berrwah, Queensland (lat. 27°S). Each hay was chopped to 1·2–2·5 cm and half was milled and pelleted. The nitrogen content of the material was increased by late nitrogen fertiliser application from 0·59 to 1·15 per cent but soluble carbohydrate was decreased from 14 to 7·9 per cent. Dry matter digestibility was not significantly affected by nitrogen fertiliser application. Although soluble carbohydrate content was decreased, protein content was increased and there was probably little change in the proportion of cell contents in the plant. On the other hand, voluntary intake was greatly increased in the high nitrogen treatment, since the nitrogen content was raised above the figure of 1·1 per cent, which is regarded as important for the maintenance of normal intake. Nitrogen apparent digestibility was also greatly increased.

Table 5.7 Fertiliser nitrogen and nutritive value of *Digitaria decumbens* hay. (*Source:* Minson, D. J. (1967). *Br. J. Nutrit.* **21**:587.)

Character	Hay Treatment	Without Nitrogen Fertiliser	With Nitrogen Fertiliser
Per cent nitrogen		0·59	1·15
Per cent soluble carbohydrate		14·0	7·9
Voluntary intake ($g/W^{0.75}$/day)V	Chopped	31	48
	Pelleted	35	62
Dry matter digestibility (per cent)	Chopped	48	52
	Pelleted	45	48
Nitrogen apparent digestibility (per cent)	Chopped	8	50
	Pelleted	12	48

Table 5.8 Fertiliser nitrogen and seasonal nutritive value of *Pennisetum clandestinum*. (*Source:* Minson, D. J. (1973). *Aust. J. Exp. Agric. Anim. Husb.* **13**:153.)

Time	**Digestibility** (per cent)		**Intake** ($g/W^{0.75}$/day)	
	Low Nitrogen	*High Nitrogen*	*Low Nitrogen*	*High Nitrogen*
December	60	63	61	61
January	63	64	68	63
February	59	54	57	50
March	57	58	63	61
April	53	59	55	67
Mean	58	59	61	61

A different situation can be seen in Table 5.8, which reports the nutritive value of 28-day regrowths of *Pennisetum clandestinum* grown at Gatton, Queensland (lat. 28°S) on a fertile 'prairie' soil developed on an alluvial flood plain. The low and high nitrogen treatments received 57 and 230 kg nitrogen per ha respectively after each 28 day cut. Over the summer–autumn period, high nitrogen increased pasture growth but did not consistently affect either dry matter digestibility or voluntary intake. In this case the low nitrogen treatment produced hays averaging 1·7 per cent nitrogen content. The lowest value was 1·5 per cent which is well above the critical value for animal protein deficiency.

Unless pastures are seriously nitrogen deficient, nitrogen fertiliser does not usually influence pasture quality either positively or negatively. However, by increasing pasture growth it enables more stock to be carried.

3 Diagnosis of mineral deficiencies and toxicities

The only reliable method of determining fertiliser needs is to measure response to fertilisers of pastures growing in the field, by experiment. However, some good guesses can be made on the basis of plant and animal visual symptoms, plant and soil analysis, and by recognising the characteristics of particular soil and vegetation associations.

(a) Visual symptoms

The identification of visual symptoms of deficiency can be straightforward for many elements. This approach is qualitative only; in other words it may indicate the nature of the problem but not the amount of fertiliser needed to rectify it. Whilst serious deficiency may produce well known symptoms which are easily recognised, farmers are often dealing with moderate insufficiencies which make the pasture grow more slowly without causing easily recognised abnormalities. Multiple deficiencies are more difficult to recognise, and some nutrient deficiencies take very different forms in different species. However, people who are very familiar with pasture plant nutrition can often recognise a problem in the field from their previous experience of growth malformations.

The first diagnostic characteristic is whether the young or the old leaves are affected. The more easily moved nutrients are readily carried to the growing points, and the older tissues first show visual deficiency symptoms in the case of nitrogen, sulphur, molybdenum, phosphorus, potassium and magnesium. On the other hand, deficiency of the less mobile nutrients shows first at the meristems and in the young leaves; this is the situation for calcium, boron, zinc, iron and to a lesser extent, copper and manganese deficiencies, and manganese toxicity. The following notes summarise common symptoms for particular mineral deficiencies; these need to be modified for some species and for the stage of development.

Nitrogen: This is the most common nutrient deficiency. Plants show a general chlorosis, starting with the older leaves; branching is reduced, and growth is retarded.

Sulphur: Similar to nitrogen deficiency, except that the whole plant is often affected more rapidly than for nitrogen. Sulphur deficient plants do not become green when nitrogen fertiliser is added.

Molybdenum: Similar to nitrogen deficiency. Legumes bear many small, ineffective nodules. The chlorosis is sometimes interveinal, and leaf margins tend to curl and roll.

Phosphorus: Phosphorus deficient plants often exhibit purple or red discolorations of the older leaves, especially on the veins, and this may extend to the stems. At first leaves are darker green or more blue-green than normal. Growth is stunted, and the deficiency is more easily seen in seedlings than in old plants.

Potassium: Interveinal necrotic spots are evident on the leaves, which also show firing of the distal leaf margins, which later causes death of the distal end of the older leaves. Terminal and lateral buds may die.

Magnesium: A mosaic pattern of chlorosis is seen on the older leaves; there is often a red or yellow discoloration from the leaf margin and a yellowing of the veins. Leaf withering also occurs. Some species, e.g. *Cenchrus ciliaris*, show brilliant red patches.

Calcium: The meristems are affected and may die back. The younger leaves are small and rolled together in legumes such as *Desmodium intortum* and *Glycine wightii*. In grasses such as *Setaria anceps* and *Paspalum plicatulum* the distal ends of the young leaves turn brown, twist, and die. Seed setting may be reduced. In *Macroptilium atropurpureum* premature fall of the older leaves may occur.

Boron: Boron deficient plants usually show damaged growing points, and short internodes. The young leaves are much reduced, and show wavy margins in *D. intortum*. In *G. wightii* the main stem below the growing point is grey. Seed formation is reduced.

Zinc: The terminal shoots often show a rosette of tiny leaves bunched together. The younger leaves are twisted and necrotic. Interveinal chlorosis is evident in *Glycine*.

Manganese: The young leaves are affected, often showing interveinal chlorosis. *G. wightii* and *S. humilis* show dark-brown spots on the upper leaf surfaces. *M. atropurpureum* has small, dish shaped leaves with withered tops. Manganese toxicity also causes interveinal chlorosis of young leaves.

Iron: The young leaves show a general chlorosis, and may be yellow or white. This chlorosis is first interveinal, and leaf margins inroll and die. *Chloris gayana* shows yellow and green stripes along the leaf. Tillering is reduced.

Copper: The young leaves first become pale to greyish green. In *D. uncinatum* and *Centrosema pubescens* marginal necrosis occurs, and the leaflets become dish shaped with rolled edges. Interveinal chlorosis is prominent in *Stylosanthes*. The young shoots die back. Seed formation is reduced.

(b) Plant analysis: 'critical' values

The content of mineral nutrient in pasture shoots shows whether or not sufficient minerals are present. As the soil nutrient supply increases, plant nutrient uptake increases, and nutrient concentration in the shoots increases as pasture yield also rises. A point is reached, often corresponding to point C on Fig. 5.2, where increasing 'luxury' nutrient concentration is no longer associated with increased pasture yield. A knowledge of the 'critical' concentration of nutrient (where adequate supply has been reached) can guide fertiliser applications. Some scientists regard the critical concentration as the concentration at which 90 per cent of maximum yield occurs. There are other definitions which lead to a similar practical result.

Table 5.9 Critical value of phosphorus and potassium adequacy and of manganese toxicity for pasture legumes. (*Sources:* (1) Andrew, C. S. and Robins, M. F. (1969). *Aust. J. Agric. Res.* **20**:665. (2) Andrew, C. S. and Robins, M. F. (1969). *Aust. J. Agric. Res.* **20**:999. (3) Andrew, C. S. and Hegarty, M. P. (1969). *Aust. J. Agric. Res.* **20**:687.)

Species	Phosphorus[1] (per cent)	Potassium[2] (per cent)	Manganese Toxicity[3] Threshold Value (ppm)
Tropical			
Centrosema pubescens	0·16	0·75	1 600
Desmodium intortum	0·22	0·72	
D. uncinatum	0·23	0·8	1 160
Glycine wightii	0·23	0·8	560
Leucaena leucocephala			550
Lotononis bainesii	0·17	0·9 ?	1 320
Macroptilium atropurpureum	0·24	0·75	810
M. lathyroides	0·20	0·75	840
Stylosanthes humilis	0·17	0·6	1 140
Vigna luteola	0·25		
Temperate			
Medicago sativa	0·24	1·2	380
M. truncatula		1·0	560
Trifolium fragiferum		1·0	510
T. repens		1·0	650

Table 5.9 shows the critical values for three nutrients for a number of tropical and temperate pasture legumes. These values are based on the dry matter content of plant tops at the 'immediate pre-flowering stage'. There are large differences between species, showing how important it is to discover the nutritional behaviour of a particular pasture species. Low values of 0·16 per cent phosphorus for *C. pubescens* and of 0·17 per cent phosphorus for *L. bainesii* and *S. humilis* show that these species use phosphorus very efficiently. A higher range of values (0·24–0·25 per cent phosphorus) was recorded for *M. atropurpureum* cv. Siratro, *Medicago sativa*, and *V. luteola*; *M. lathyroides, D. intortum, D. uncinatum* and *G. wightii* were intermediate. Other work at the CSIRO Cunningham Laboratory has given the following critical phosphorus percentage values for tropical grasses: *Digitaria decumbens* 0·16; *Melinis minutiflora* 0·18; *Panicum maximum* 0·19; *Sorghum almum* 0·20; *Pennisetum clandestinum, Setaria anceps* 0·22; *Chloris gayana* 0·23; *Paspalum dilatatum* 0·25; *Cenchrus ciliaris* 0·26.

The critical potassium values in Table 5.9 show a range from 0·6 per cent for *S. humilis* to 1·2 per cent for *M. sativa*. The values for the temperate legumes were higher than for the tropical legumes tested.

The concept of critical values can be inverted to decide whether mineral toxicity is present. The manganese toxicity threshold values in Table 5.9 indicate the manganese concentration in plant tops corresponding to the yield point which is 5 per cent lower than the maximum. The temperate legumes tested and the tropical legumes *G. wightii* and *L. leucocephala* all showed susceptibility to manganese toxicity, with threshold values less than 600 ppm. On the other hand *L. bainesii* and *C. pubescens* recorded values of 1 320 and 1 600 ppm. This explains why they have a good field performance on very acid soils with high manganese availability.

There are many difficulties in developing a working application of this knowledge in the field. Critical values are obtained for conditions where yield response is not limited by shortage or toxicity of any other nutrients. This would be an unusual field situation. The critical value decreases as the plant ages, and it is difficult to arrange field sampling at comparable stages. Nutrient concentration in the whole plant tops is sensitive to changes in the ratio of leaf to stem, and it may be more suitable to base critical values on leaves which are all at the same stage of expansion, or on young shoots all of the same size. For example the shoots could be taken for testing, using the part of the stem from the growing point to the third fully expanded leaf. Finally, under dry conditions critical values are decreased.

(c) Soil analysis

Soil analysis is less reliable than plant analysis for deciding fertiliser application levels, and intelligence and skill are required to interpret the results of soil tests. It is necessary to estimate the response of the pasture to the soil situation. Nutrient uptake by the pasture is not simply related to nutrient content of the soil, since many factors interfere with nutrient availability to plant roots. Many soil samples must be taken to fully represent the field in question and the rooting medium of the pasture. For example, the surface soil may be acutely phosphorus deficient but abundant phosphorus may be present at depth.

Soil pH is the first guide to mineral availability and the possibility of nodulation problems in legumes. Organic carbon and total nitrogen content of virgin soils also give a general indication of the likelihood of fertility problems. The total content of other nutrients is a very rough guide, and special extractants have been developed to indicate availability to plants. Comparing these tests with field pasture response in a particular region is a much more reliable way of deciding which fertiliser is needed. For instance, work by the Queensland Department of Primary Industries in coastal north Queensland compared the response of tropical pastures to superphosphate with 'available' soil phosphorus assessed by different methods. Soil phosphorus status was estimated by the Kerr-von Stieglitz method (0·01N sulphuric acid extraction), the Bray P_2 method (0·1N hydrochloric acid/0·03N ammonium fluoride extractant) and the Colwell method (sodium bicarbonate extractant). The Kerr-von Stieglitz and Colwell methods extracted similar amounts of phosphorus over a wide range of soils, compared well with results from field tests of pasture response, assessed as 'per cent yield (= yield without fertiliser/maximum yield with fertiliser × 100). The critical soil values giving 90 per cent yields were 17·5 ppm phosphorus for the Kerr-von Stieglitz method, and 20 ppm phosphorus for the Colwell method. The Bray P_2 method was satisfactory on some soils, but gave results which compared poorly with phosphorus response on basaltic derived soils, whose high content of iron (III) oxide may have interfered with the method.

Information on exchangeable cations and chloride content is a useful diagnostic guide.

(d) Soil and vegetation groups

Soil type, especially soil origin, and the vegetation association it bears, can be reliable guides to pasture establishment fertiliser needs. They are particularly useful as guides in pasture development operations with virgin land little altered by previous fertiliser application or agricultural use.

These principles are well illustrated by a series of studies carried out by the Queensland Department of Primary Industries to characterise land suitable for pasture development in coastal Queensland. A survey of the region was followed by a very wide series of fertiliser experiments. It was found that the results of the experiments could be classified readily according to the origin of the soil in question, and the vegetation growing on it. Table 5.10 lists the major soil groups, and the nutrient deficiencies recorded on them. These varied from a single phosphorus deficiency on mixed alluvials containing basaltic derivatives to multiple deficiencies on inland beach sands of phosphorus, potassium, calcium, sulphur, copper, zinc, molybdenum and boron.

Further details of the variation in fertility were obtained from a study of the vegetation associations. On well-

Table 5.10 Fertility of soil and vegetation associations in north Queensland. (*Source:* Teitzel, J. K. (1972). *Qld Dept. Prim. Indus. Div. Pl. Ind. Adv. Leaflet* 1139.)

Soil Type	Mineral Deficiencies
Mixed alluvial containing basalt	phosphorus
Basaltic	phosphorus, calcium, sulphur, molybdenum
Metamorphic	phosphorus, potassium, calcium, sulphur, molybdenum
Mixed alluvial with little basalt	phosphorus, potassium, calcium, copper
Granitic	phosphorus, potassium, calcium, sulphur, copper, zinc
Beach sand	phosphorus, potassium, calcium, sulphur, copper, zinc, molybdenum, boron

Vegetation Type	**Fertiliser Planting Recommendation on Granite Soils** (kg/ha)
Rain forest	250 single superphosphate
Bastard scrub	250 superphosphate 60 potassium chloride
Open forest	500 superphosphate 60–120 potassium chloride 8 copper sulphate 8 zinc sulphate
Grassy woodland	500 superphosphate 120 potassium chloride 8 copper sulphate 8 zinc sulphate
Palm swamp	250 superphosphate
Narrow leaf tea-tree	500 superphosphate 60–120 potassium chloride 8 zinc sulphate
Broad leaf tea-tree	500 superphosphate 120 potassium chloride 8 zinc sulphate

drained sites, rain-forest occurs on the most fertile soils. In such cases a closed canopy is formed by many different types of trees. Mixed sclerophyll-mesophyll vine forest or 'bastard scrub' is the next member of the unit; some large acacias, eucalypts, and *Tristania suaveolens* emerge above a mesophyll understory. On still less fertile soils an open forest formation of these and other trees occurs, but with a thin ground layer. Grassy woodland represents the least fertile site; the trees are smaller and sparser and there is a well developed grass ground layer dominated by species of *Heteropogon, Themeda, Imperata*, and *Panicum*. On poorly-drained sites, a palm swamp indicates the most fertile soil; this is characterised by *Archontophoenix* or *Licuala* with *Calamus* in the understory. On less fertile soils, a narrow-leaf tea-tree (*Melaleuca*), tall swamp-forest develops, whilst on the least fertile unit a broad-leaf tea-tree, low layered swamp-forest occurs. The lower half of Table 5.10 gives fertiliser recommendations for establishing legume based pastures on each of these vegetation units when occurring on granitic soils. This table indicates the variation in fertility for both well-drained and poorly-drained units. Not all vegetation units occur on all of the soils listed; for instance, no rain-forest or bastard scrub occurs on beach sands. Fertiliser recommendations have been worked out for the soil/vegetation combinations present, and these have been of great assistance to producers developing these lands for beef production.

4 Pasture fertiliser practice

Fertiliser practice is treated in depth in a companion book in this series, 'Tropical soils and fertiliser usage'. In this section mention is made of some aspects which are especially pertinent to tropical pastures regarding frequency and timing of fertiliser application, and nutrient sources. Fertiliser placement was discussed in the last chapter about pasture establishment.

The danger of environmental pollution from using too much fertiliser should always be remembered. Leaching of nitrate from pasture and crop land may lead to abundant growth of filamentous algae in streams and reservoirs. These die and accumulate on the bottom of the stream. The breakdown of this organic matter by decomposing bacteria consumes the available oxygen in the water, killing fish and other marine life. High nitrate content in the water supply for humans or for crops intended for human consumption can be harmful. This problem can be avoided by the use of sensible nitrogen fertiliser levels, which do not cause a leaching of nitrogen, and by consuming surplus nitrogen with grass pastures or with crops with high nitrogen demand.

(a) Frequency and timing of fertiliser application

It is usually thought to be desirable to apply fertiliser heavily in the initial pasture establishment season, in order to create a bank of soil fertility. Heavy fertilising of legumes is directed to creating legume dominance and increasing the amount of nitrogen fixation upon which subsequent grass growths depends. The purpose of heavy fertilising of young pastures, either legume based or grass based, is to cause a rapid build up of soil organic matter to create the exchange complex which will retain subsequent fertiliser applications and improve moisture relations, to control weeds and erosion, and to grow feed for early grazing. Later maintenance fertiliser dressings are then adjusted to maintain nutrient adequacy against the losses taking place from the system.

This heavy initial capital investment in fertiliser has proved a better plan for many pasture situations than spreading the same investment over a number of years. It is most useful for low fertility soils, where legumes work well and where the base level of production is otherwise so low that it is difficult to make a profit from the pasture. The effects of high fertiliser levels on nitrogen fixation and on botanical composition are discussed later in this section. This plan has been tested for *S. humilis* based pastures on a sandy loam at the CSIRO Pastoral Laboratory, Lansdown, Queensland (lat. 19°S). Heavy initial application of superphosphate gave a higher total pasture production over three years than split applications of the same total amount; for instance, 375 kg per ha superphosphate in the first year only was better than 125 kg per ha in each of three years.

This policy has also been applied to grass-only pastures fertilised with nitrogen. At Frankenwald, in the South African Highveld (lat. 26°S, 1 500 m altitude, 750 mm rainfall) on *Eragrostis curvula* pastures, a heavy initial dressing of nitrogenous fertiliser is made, and subsequent maintenance dressings are calculated on the basis of 80 per cent nitrogen return by the grazing animal, and 50 per cent nitrogen recovery in shoots in the next two years after application. A typical fertilising sequence would be; 120, 20, 20, 30, 30, kg nitrogen per ha in successive years.

Figure 5.5 summarises data from a grazing experiment at Fazenda Jangada, Aracatuba (1 150 mm rainfall) in the Sao Paulo State, Brazil. *Panicum maximum* pastures grown on sandy loams derived from sandstone were fertilised at the rate of 200 kg nitrogen per ha per annum for the seven years (1957–64). Productivity was estimated from live-weight gains as the Total Digestible Nutrients obtained from the pasture and are presented relative to the control unfertilised pasture production. The experiment was continued after nitrogen fertiliser application ceased in 1964, to measure the residual effects. Figure 5.5 shows that these effects were very pronounced; productivity was well maintained in 1965, and then decreased about 50 per cent in each of the next two years. The response decreased by an average of one-third each year, suggesting that nitrogen was being recycled very efficiently under grazing.

However, residual (left-over) effects are not always as great as in the two cases illustrated above, e.g. CSIRO studies at Samford, Queensland (lat. 27°S) have found little residual value of nitrogen fertiliser in years after nitrogen application, and losses from this system must be much greater.

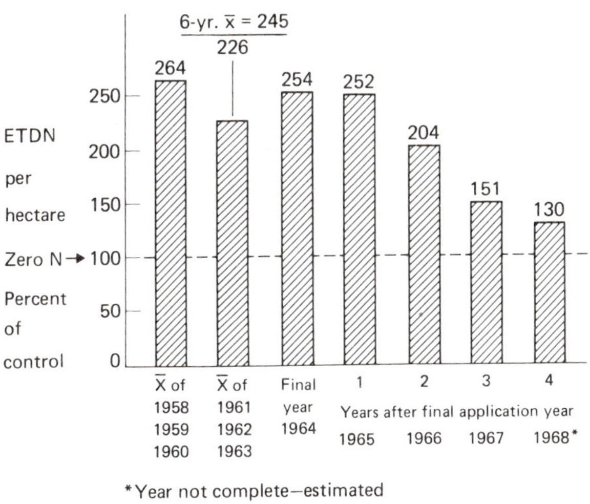

fig. 5.5 Residual effect of nitrogen fertiliser upon the yield of estimated total digestible nutrients (ETDN) per ha. (*Source:* Mott, G. O.; Quinn, L. R. and Bisschoff, W. V. A. (1970). *Proc. XI Internat. Grassld. Congr.,* Surfers Paradise p. 414.)

Attempting to build a bank of fertility quickly is of no use if the soil reacts with fertiliser to make nutrients unavailable (as occurs, for example with phosphorus fixing soils high in aluminium), or if losses from leaching or denitrification are high. The mobile nutrient potassium is very easily leached. Sulphur is another nutrient which requires frequent application. Many graziers applying single superphosphate (which contains about 40 per cent calcium sulphate) to meet both phosphorus and sulphur deficiencies have found after some years that continued annual maintenance dressings of sulphur were required, whereas phosphorus needs were much reduced. Micronutrients such as copper or manganese are usually applied in the hope that the next dressing need not be made for six or seven years. Studies on a prairie-like soil at Rodds Bay, Queensland (lat. 24°S, 810 mm rainfall) showed that molybdenum applications were needed every three to four years.

The main seasonal fertiliser application is usually made at the beginning of the growing season. The question of applying fertiliser at other times when there is not enough pasture available is mentioned in Part 6. Frequent applications of the very mobile nutrient nitrogen is desirable; for instance, a study by the CSIRO at Beewah (lat. 27°S 1 630 mm rainfall) on *Digitaria decumbens* pastures showed that liveweight gain over a three year period average 1 170 kg per ha per annum when nitrogen was applied three times per annum, and 1 440 kg per ha per annum when applied six times per annum. Fertiliser applied when pasture growth potential is less gives a smaller level of response. For instance 40 kg additional dry matter may be produced from 1 kg fertiliser nitrogen applied at the beginning of the growing season, and only 20 kg dry matter be produced from 1 kg nitrogen applied towards the end of the growing season. However, additional out of season growth is certain to be consumed, and is produced at a time when forage has a great value. Before farmers used much fertiliser nitrogen it was often thought that cool weather or dry conditions limited pasture growth in the autumn. It is now realised that nitrogen availability is often the first limiting factor and that better use of the moisture available occurs if the nitrogen limitation is removed.

(b) Nutrient sources

Access to particular fertilisers depends upon a host of

Table 5.11 Source of nitrogen and fertiliser nitrogen recovery and soil effects. (*Source:* Figarella, J., Abruna, F. and Vicente-Chandler, J. (1972). *J. Agric. Univ. Puerto Rico* **56**:410.)

Nitrogen Source	Recovery of Fertiliser Nitrogen (per cent)	Soil pH	Exchangeable Calcium and Magnesium (mol × 10^{-3}/100 g soil)
No nitrogen applied	—	5·5	18·6
Urea	43·8	4·8	14·4
Ammonium sulphate	52·9	4·5	10·2
Ammonium nitrate	46·2	4·9	14·6
Urea plus calcium carbonate	47·9	5·2	20·4
Ammonium nitrate – lime	47·8	5·5	22·0

geographical, commercial and political factors. Cost per unit of nutrient is the first consideration when choosing between alternative fertilisers. Better availability should also be considered (e.g. superphosphate rather than rock phosphate) and high nutrient content, especially if transport or distribution costs are heavy. Care should be taken when assessing fertilisers to ensure that the whole composition is considered. Farmers purchasing superphosphate need to know whether the calcium and sulphur content is of significance to them. Some sources of phosphate contain useful impurities such as zinc which may be satisfying a nutrient deficiency; a change of source may cause unexpected pasture growth problems. High analysis fertilisers have reduced transport costs. However, they have also caused real losses of pasture production where the omission of sulphur in a new product has been out of tune with soil needs, or where in some countries phosphorus is only readily sold in combination with nitrogen. This has proved disastrous to legume survival in mixed pastures, and grass dominance has resulted.

Many studies have been made of the effectiveness of various nitrogen sources on pasture. Table 5.11 compares the recovery in *Digitaria decumbens* shoots of urea, ammonium sulphate, ammonium nitrate, urea plus calcium carbonate, and ammonium nitrate-lime. This experiment was carried out over a three-year period on a deep red clay near Orocovis (600 m altitude), Puerto Rico, using annual nitrogen rates varying from 190 to 570 kg per ha. Ammonium sulphate gave the best recovery (52·9 per cent) and urea gave the poorest recovery (43·8 per cent). Pasture studies in Georgia, Alabama, Florida, Queensland, Rhodesia and other places generally support this finding

fig. 5.6 Dense, leafy growth of *Digitaria decumbens* receiving 336 kg nitrogen per ha per annum at Parada, north Queensland

that urea is a less effective provider of nitrogen to pasture, and price evaluation of urea should take account of this. Table 5.11 also shows that ammonium sulphate, urea, and ammonium nitrate application decreased soil pH from 5·5 to 4·5, 4·8 and 4·9 respectively. Exchangeable calcium and magnesium also suffered decreases in these treatments. Heavy use of most nitrogen fertiliser sources creates a need to ensure that soil acidity is not increased to dangerous levels. Ammonium sulphate is particularly damaging in this respect, and the added cost of liming should be noted when evaluating this fertiliser.

5 Nutrient supply and pasture botanical composition

The sensitivity of pasture botanical composition to nutrient supply has often been mentioned in this book. In Parts 2 and 3 the basic philosophy of choosing pasture species appropriate to the farm fertility situation and/or capable of responsiveness to improved nutrition was described. Table 3.9 indicates how the dominance of *Pennisetum clandestinum* or *Axonopus affinis* is determined by nitrogen level. In Part 4 Table 4.9 illustrated how the successful dominance of *Triodia pungens* by *Cenchrus ciliaris* was dependent on nutrient supply. Figure 4.11 showed the adverse effect of nitrogen fertiliser on *Stylosanthes guyanensis* when associated with *Oryza sativa*. In this chapter some characterisation of species with respect to nutrient requirements has been made.

The outcome of competition between species for a nutrient depends firstly upon factors controlling nutrient absorption. Species with a larger or deeper root system are at an advantage. Differences exist in the capacity of species to extract nutrients at low concentration, e.g. *S. humilis* has superior phosphorus absorption. Legume species with higher root cation exchange capacity may extract phosphorus from rock phosphate more successfully than grasses with low cation exchange capacity. Factors determining the efficiency of use within the plant are the other consideration. These include genetic capacity for high growth rate in a particular environment, and capacity for mineral retention and recirculation within the plant, e.g. some species are more successful than others in exporting phosphorus from ageing leaves to young leaves to be used again. The balance of these factors will then determine the share a species obtains of other growth factors in the environment, e.g. a species advantage derived from superior phosphorus uptake and use may then be reinforced by that species also absorbing more of the available radiation and the available moisture.

Fertiliser practice is often directed to controlling the grass/legume balance. Under conditions of low nitrogen supply it is usual for the dominance of nodulated legumes to be encouraged by application of non-nitrogen fertilisers. Figure 5.7 shows the first season yield of five pasture mixtures grown at the University of Queensland Mt. Cotton Farm (lat. 28°S) on a red-yellow podzolic soil. After cultivation, an unsown control natural pasture dominated by the grasses *Axonopus affinis, Digitaria adscendens*, and *Eragrostis* spp. and the sedges *Cyperus polystachyos, C. sesquiflorus* and *Fimbristylis dichotoma* was allowed to volunteer, and four other pastures were sown to *Setaria anceps* cv. Kazungula with either *Desmodium intortum, Lotononis bainesii, Stylosanthes guyanensis*, or *Macroptilium atropurpureum* cv. Siratro. Calcium dihydrogen phosphate was applied at rates of 0, 11, 22, 45 and 89 kg phosphorus per ha.

Figure 5.7 shows that (with the exception of *Lotononis*) increasing phosphorus level increased legume yield as represented by the top white bars in the diagram. Yield of the unfertilised pastures was about 2 200 kg per ha. Yield increased to a maximum of 8 300 kg per ha at the 89 kg phosphorus per ha level in the *Desmodium* pasture, of which *Desmodium* was 58 per cent of the yield. Yield of *Setaria* increased to the 45 kg phosphorus per ha level, but was dominated by the legumes at the highest level. The unsown control volunteer pasture (represented by the bars marked C in the figure) increased to 3 500 kg per ha yield at the 22 kg phosphorus per ha level, but did not increase in yield at higher levels, indicating the inability of these species to perform well under conditions of high phosphorus supply. The proportion of sedges and volunteer grasses decreased in the sown pastures where phosphorus was applied.

The various species in this study showed to advantage in different ways. Where competition for phosphorus was most serious in the unfertilised treatment, efficiency of phosphorus use (dry matter yield per unit of phosphorus uptake) was high in *Setaria*. The superior performance of *Desmodium* at the highest phosphorus level arose both from its efficient use of phosphorus and its high phosphorus uptake. At low phosphorus levels its phosphorus uptake was much less. The good performance of *Lotononis* at low phosphorus supply was a result of its superior phosphorus uptake. Both phosphorus uptake and efficiency of use of phosphorus were high in *Stylosanthes* in low phosphorus treatments.

Many other studies have indicated that good phosphorus supply aids early legume dominance. This has been shown, for example, in *Panicum maximum*/*Dolichos argentis* pastures on sandy soils under coconuts on the Tanzania coast, where applications of phosphorus and potassium have increased the legume content, and nitrogen fertiliser applications have decreased it. When the nitrogen of the soil has been improved, it is usual for grass

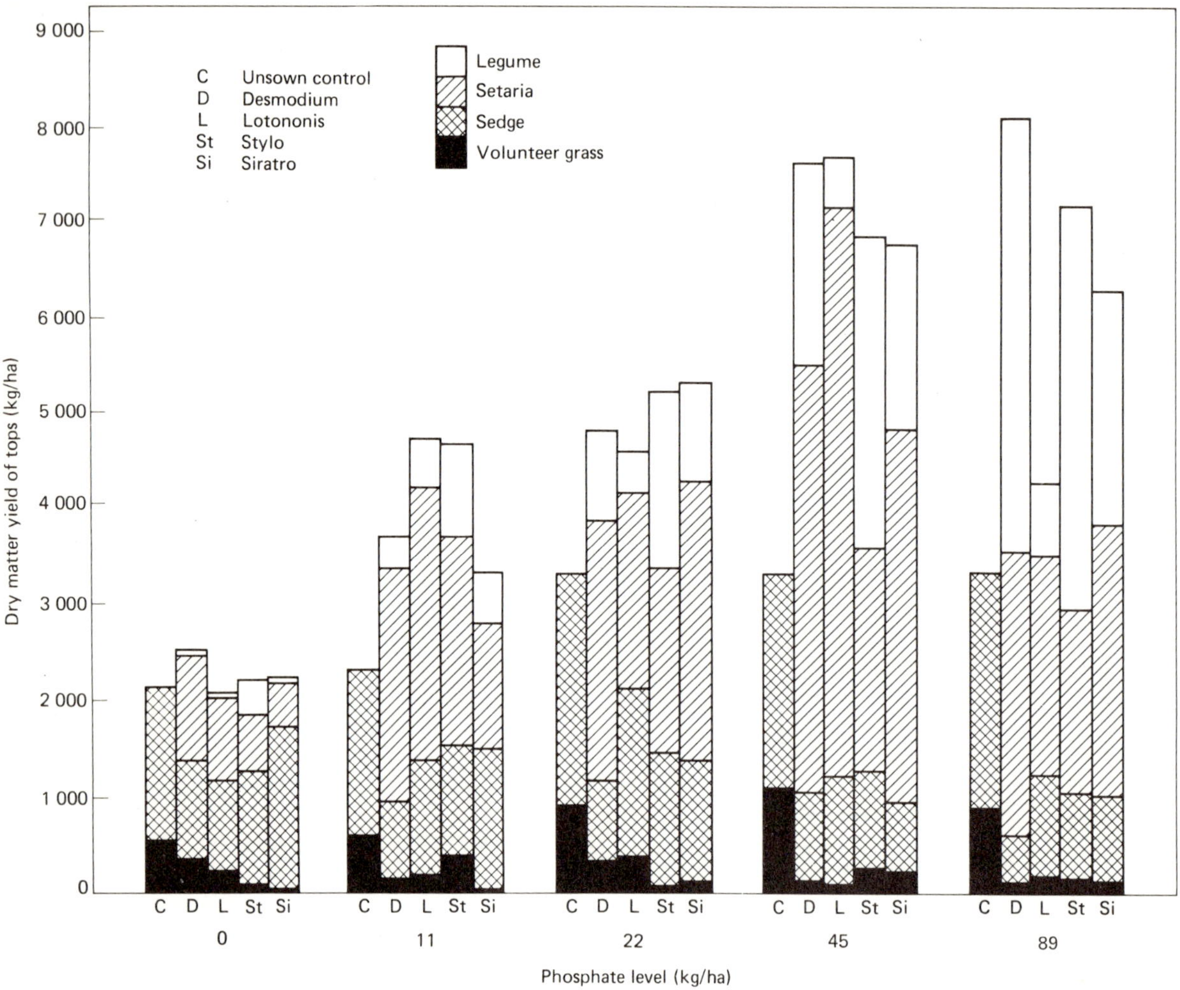

fig. 5.7 Effect of phosphate application on the yield of legume, sown grass, sedge and volunteer grasses for different pasture mixtures. (*Source:* Blunt, C. G. and Humphreys, L. R. (1970). *Aust. J. Exp. Agric. Anim. Husb.* **10**:431)

fig. 5.8 *S. guyanensis* and *Setaria anceps* on a red-yellow podzolic soil at Mt. Cotton, south-east Queensland. Left-hand side: 89 kg phosphorus per ha; right-hand side: no added phosphorus

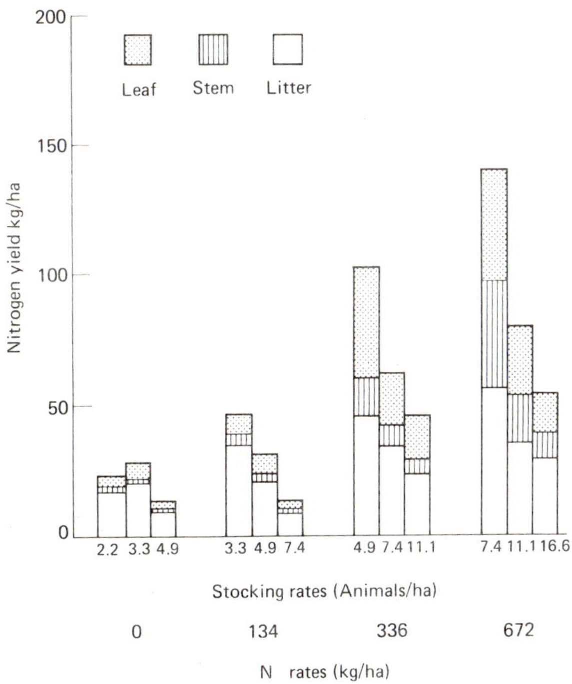

fig. 5.9 Effect of increasing nitrogen rates and stocking rates on the size of the nitrogen pools of leaf, stem and litter in *Pennisetum clandestinum* pastures. (*Source:* Mears, P. T. and Humphreys, L. R. (1974). *J. Agric. Sci. Camb.* **83**:451)

dominance to follow. On soils sufficiently high in nitrogen at the time of pasture establishment, high phosphorus fertiliser application rates give grass dominance from the start of the pasture.

The application of nitrogen fertiliser almost universally decreases legume content. This is not because of any harmful effect on the legume itself; legumes respond well to nitrogen fertiliser if growing as pure stands. The effect is due to the higher growth rates of grasses when their growth potential is not limited by nitrogen deficiency. This usually results in shading of the legume or in some other growth factor, e.g. moisture, being denied to the legume by the more vigorous grass growth.

Potassium deficiency is usually more serious in legumes than in grasses. This is most easily noticed in mixed culture, where grass roots are more efficient than legume roots in extracting potassium from the soil. Table 5.12 shows the result of an experiment where *Desmodium* and *Setaria* were grown in narrow soil panels set side by side, so that their tops were intermingled. Their roots were either associated in the same panels, or were separated in different panels. When no potassium was added, *Desmodium* yield was sharply reduced (8·9 g) if the tops and roots of the two species were associated and was much improved (14·0 g) if the tops associated and the roots separated. This effect disappeared if potassium was in good supply. Grass dominance in many pastures is due to insufficient use of potassium fertiliser.

Table 5.12 Potassium supply and competition between *Setaria anceps* and *Desmodium intortum*. (*Source:* Hall, R. L. (1971). *Aust. J. Exp. Agric. Anim. Husb.* **11**:415.)

Treatment	**Plant Yield** (g)	
	Desmodium	***Setaria***
Tops and roots associated		
No added potassium	8·9	17·3
Added potassium	16·2	16·0
Tops associated, roots separated		
No added potassium	14·0	12·2
Added potassium	16·8	14·5

6 The nutrient cycle in grazed pastures

Nutrients move within the cycle of the soil–pasture–animal system. Intelligent grazing and fertiliser management can accelerate the processes within the cycle to maximise off-take of animal products, to minimise other nutrient losses from the cycle, and to conserve and add to the nutrient pools which affect production. The major nutrient pools are:

(a) the soil, which has inorganic nutrient sources available to plants and in process of decomposition, and a bank of organic matter in varying stages of breakdown and availability;

(b) the living plant tissues, subject to senescence, to physical erosion, and to consumption above and below ground by organisms of varying level;

(c) the animals, comprising the body tissues and alimentary tract of domestic animals, and the natural fauna ranging from ungulates down to arthropods and bacteria.

The gains or transfer to the soil–pasture–animal cycle come from fertiliser, from imported feed stuffs, from industrial waste and from atmospheric nitrogen converted to available forms by electrical storms, by free-living nitrogen fixing organisms or by *Rhizobium* in legume root nodules. Gains can also come from decomposition of rock. The losses or transfers from the grazing system are due to sale of animal products, leaching, erosion, and gaseous losses, e.g. denitrification and ammonia loss from urine.

A nutrient cycle has many ways of operating. The channels along which a nutrient moves can have differing importance depending on the way the pasture is managed. The output of animal product is not fixed by the size of the nutrient pools, but by the rates of flow between the various pools.

The fertility which animals return to the pasture as they graze accounts for most of the nutrients ingested. The growing sheep retains only about 4 per cent of minerals and nitrogen whilst the milking cow retains about 10 per cent of minerals and 25 per cent of nitrogen eaten from the pasture. It follows that night-yarding or housing of animals causes a rapid transfer of nutrients from the pasture, unless these are carefully collected and returned. Manure may be collected for sale or for use on crop land; this fertility loss from the pasture land should be taken into account when deciding the success of the operation. Transfer of nutrients from point to point also occurs within a grazed field, and fertility gradients occur from stock shelter and camping places, watering points, and gateways. Urine and dung patches are very unevenly distributed, and each represents a luxury supply of nutrients, e.g. a sheep urine patch may have received the equivalent of 480 kg nitrogen per ha in one dressing. The highest concentration of nutrients in a camping place may occur on bare ground, where the pasture has been eaten out; this is surrounded by a zone of luxury availability, often grading out to deficient conditions. Movement of shelter and watering points reduces this problem.

fig. 5.10 Hereford steers grazing well fertilised *D. decumbens* at Beerwah in the coastal lowlands of southern Queensland

Examination of dung and of urine patches can assist diagnosis of mineral deficiences. Most of the potassium and boron voided are contained in the urine, whilst almost all the phosphorus (organic and inorganic), calcium and most of the magnesium and the micronutrients iron, manganese, zinc, copper and cobalt are voided in the faeces. Nitrogen and sulphur are voided in the faeces at rates of about 0·8 and 0·1 g respectively per 100 g pasture dry matter eaten, the balance appearing in the urine. Much urinary nitrogen is lost to the air from the herbage and soil, amounting to perhaps 12 per cent in a dry summer. The presence of coprophagous beetles, rapidly burying faeces, greatly helps the nitrogen economy of the system.

Figure 5.9 shows the size of the above-ground nitrogen pools in *Pennisetum clandestinum* pastures at Lismore, N.S.W., Australia (lat. 29°S, 1 700 mm rainfall) in June, 1970. These pastures received ammonium nitrate at rates of 0, 134, 336 and 672 kg nitrogen per ha, and were continuously grazed at rates varying from 2·2 to 16·6 yearlings per ha. The productivity of these pastures is discussed in the next chapter, and at this point attention is drawn to the way in which nitrogen yield of the feed on offer is increased with the nitrogen fertiliser input, and decreases as stocking rate increases. For the most leniently grazed treatment of each nitrogen fertiliser level, total nitrogen in above ground plant parts increased from 20 kg nitrogen per ha in the control treatment to 140 kg nitrogen per ha in the N_{672} treatment, and heavy grazing pressure reduced the size of the nitrogen pool at each fertiliser level. Heavy grazing reduced the litter nitrogen pool and usually increased the leaf nitrogen: stem nitrogen ratio. The litter, which was composed of dead leaves, surface litter and fragments of faeces, represents an important reserve of nutrients, and an energy source for earthworms, arthropods, gastropods, fungi and bacteria. Decomposition is assisted by heavy grazing and nitrogen fertiliser, since these narrow the carbon:nitrogen ratio of the litter. The reduced size of the pool of litter and other organic matter under heavy grazing increases the recirculation of nutrients in the soil–pasture–animal system.

7 The legume-*Rhizobium* symbiosis

Tropical pasture legumes fix as much nitrogen as temperate pasture legumes. There have now been many careful studies carried out in a number of tropical countries to measure this. There is no excuse for scientists to attempt to adapt temperate legumes to unsuitable tropical situations in the hope that they will produce more nitrogen than tropical legumes. For instance, *Centrosema pubescens*/*Cynodon plectastochyus* pastures in Nigeria fixed an estimated 280 kg nitrogen per ha per annum. In Kenya, *Glycine wightii* added 180 kg nitrogen per ha per annum for the first five years after sowing, and 110 kg per annum in the next four years. *C. pubescens* added 235 kg nitrogen per ha in five months growth in west Malaysia. On the north Queensland wet coast *C. pubescens*/*Panicum maximum* pastures after 16 years had higher nitrogen content in the top 15 cm soil than pasture of *P. maximum* alone, equivalent to annual nitrogen fixation of 103 kg per ha. Over a three-year period, *C. pubescens*/*Pennisetum purpureum* pastures at the same location added 138 kg nitrogen per ha per annum; *Pueraria phaseoloides*/*P. purpureum* pastures added 145 kg nitrogen per ha per annum. In south-eastern Queensland, *Macroptilium atropurpureum* with *Paspalum plicatulum* gave a nitrogen yield equivalent to adding 192 kg nitrogen fertiliser per ha per annum to a sward of *P. plicatulum* only, and a dry matter yield equivalent to adding 101 kg nitrogen fertiliser per ha per annum. At Shika in Nigeria, *Stylosanthes guyanensis* with *Chloris gayana* gave a nitrogen yield equivalent to adding 187 kg nitrogen fertiliser per ha per annum to a pasture of *C. gayana* only, and a dry matter yield equivalent to adding 84 kg nitrogen fertiliser per ha per annum. These levels of nitrogen production are very high, even if they do not allow companion tropical grasses to grow to their maximum level.

The first stage in ensuring effective nitrogen fixation is a compatible host-*Rhizobium* relationship. Rhizobial specificity and legume establishment techniques were discussed in Part 4. Table 4.7 lists the inoculation requirements of the main pasture legumes. The tropical legumes are older on the evolutionary scale and more numerous than the temperate legumes, which are generally more specialised in their affinities with *Rhizobium* and less tolerant of low pH situations. However, even within the cowpea-type promiscuous group there are very real differences in the effectiveness of particular rhizobial strains, and the selection of host-*Rhizobium* combinations which fix more nitrogen is advantageous.

(a) Weather effects on nitrogen fixation

The amount of nitrogen fixation increases with the amount of legume growth as indicated in Part 3. It follows that weather conditions which give high legume yields promote more nitrogen fixation. In dry environments, much less nitrogen fixation is possible. A CSIRO study of continuously grazed *Stylosanthes humilis* pastures at Rodds Bay, Queensland (lat. 24°S, 810 mm rainfall) showed that the change in soil nitrogen varied with the season. In wet years, a gain of up to 30 kg nitrogen per ha per annum was recorded, whilst in dry years a net loss of nitrogen of as much as 29 kg per ha might occur. This was probably associated with volatilisation of nitrogen from dung and urine, and with decreased growth of *S. humilis* in a dry year.

Studies in Malaysia have shown better nodulation on wet soils, at least until the plants were actually flooded. Malaysian studies have also shown the benefits of organic manures on nodulation. Both these effects may be related to an important factor controlling nodule initiation: the inhibiting effect of soil nitrate. Experiments in East Africa discovered that the amount of soil nitrate released upon soil wetting was proportional to the length and the severity of the dry period before wetting. It may be expected that nodulation on some soils high in nitrogen will be delayed if sowings are made on rain following a long dry season, since much soil nitrate will be mobilised. Soils constantly wet will contain less nitrate, and organic manures may be expected to consume available nitrate for further decomposition; under both these conditions better nodulation is expected.

Plants shed their nodules in response to cold, drought, flooding, and shading.

(b) Effects of mineral nutrition on nitrogen fixation

Table 3.12 illustrated the effects of soil type on nodulation and nitrogen fixation of *Glycine wightii*. In this chapter the benefits of fertilising the legume in order to fix more nitrogen have been emphasised. Special attention should be given to the nutrients specifically involved in nitrogen fixation, molybdenum, iron and cobalt, and to the nutrients which particularly favour legume growth, such as phosphorus, potassium, sulphur and calcium. One study on a latosol in the humid lowlands of Hawaii showed that the nitrogen yield of a *Desmodium intortum/Digitaria decumbens* pasture varied from 40 kg per ha per annum where no phosphorus was added, to 410 kg nitrogen per ha per annum where very high rates of phosphorus fertiliser were applied. On the Queensland coastal lowlands at Beerwah, the estimated annual nitrogen accumulation of *D. uncinatum* pasture was 71 kg per ha where 520 kg single superphosphate per ha was applied over six years. Nitrogen accumulation increased to 126 kg per ha per annum where 4 100 kg single superphosphate was applied over six years.

Table 5.13 is drawn from the experiment already discussed with Figure 5.7, which showed the response to added phosphorus of four legume pastures each grown with *Setaria anceps* and of a volunteer natural pasture. This table shows the yield response in terms of additional nitrogen in the mixed pasture as a result of the phosphorus fertiliser applied. For *Desmodium intortum*, the efficiency of response rose to 1·04 kg additional nitrogen per kg phosphorus applied at the P_{45} and P_{89} level. *Lotononis bainesii* was efficient at low phosphorus level, and high values (0·88 to 1·74 kg nitrogen per kg phosphorus) were recorded by *Macroptilium atropurpureum* cv. Siratro and *Stylosanthes guyanensis* in the range 11 to 45 kg phosphorus applied per ha. Clearly, when deciding if the cost of phosphorus fertiliser is worthwhile one must consider the secondary nitrogen benefits, as well as in the conventional terms of dry matter yield response and phosphorus recovery. These benefits can result in increases in stock production. For instance a CSIRO study at Rodds Bay, Queensland (lat. 24°S, 810 mm rainfall) on a *S. humilis* pastures showed that over seven years liveweight gain increased from 93 kg per ha per annum on an unfertilised pasture to 148 kg per ha per annum on one fertilised with superphosphate, potassium and molybdenum.

Table 5.13 Phosphorus supply and nitrogen yield of mixed pastures at Mt. Cotton south-eastern Queensland. (*Source:* Blunt, C. G. and Humphreys, L. R. (1970). *Aust. J. Exp. Agric. Anim. Husb.* **10**:431.)

Pasture	**Nitrogen per kg Phosphorus** *Nitrogen Yield Response at Four Levels of Phosphorus* (kg/ha)			
	11	**22**	**45**	**89**
Volunteer	–0·11	0·25	0·13	0·04
***Setaria anceps* with:**				
Desmodium intortum	0·19	0·61	1·04	1·04
Lotononis bainesii	1·85	0·77	0·97	0·24
Macroptilium atropurpureum	1·74	1·51	1·60	0·83
Stylosanthes guyanensis	0·88	1·26	0·94	0·63

Nitrogen fixation is most efficient on soils low in nitrogen but on which all other nutrient deficiencies are corrected. In pure legume or legume dominant swards, nitrogen fixation is reduced under conditions of good soil nitrogen supply, and the legume takes up available nitrogen from the soil rather than fix nitrogen from the atmosphere. One CSIRO study on a podzolic soil at Samford, Queensland (lat. 27°S, 1 150 mm rainfall) showed that growing *M. atropurpureum* increased total nitrogen from 0·089 per cent to 0·125 per cent. However, on a more fertile latosol with an initial total nitrogen content of 0·167 per cent no increase in soil nitrogen occurred after growing the legume. Grass competition is also more serious on soils with good nitrogen supply, and some of the most successful examples of legume adaptation occur on poor sands of low nitrogen content.

(c) Management effects on nitrogen fixation and transfer

The adverse effects of severe defoliation on nodulation are mentioned in Part 4. Nodules are the organs of the plant first affected by a carbohydrate shortage induced by grazing or cutting. On the other hand, the most effective mechanism of transfer of nitrogen from legume to grass is via the grazing animal consuming legume shoots and voiding most of the nitrogen in relatively available forms. A balance therefore has to be found between defoliation which will not imperil legume growth and nitrogen fixation too greatly, and defoliation which will transfer nitrogen to companion grasses. This problem is less evident for low growing legumes, e.g. *S. humilis*, which are more resistant to heavy grazing and which benefit from the consumption of taller companion grasses.

Most of the nitrogen fixed by legumes enters the shoots. This proportion varies from 55 to 80 per cent, according to severity of defoliation. Some transfer of nitrogen from root decay is feasible, especially with annual species. The contribution caused directly by nodule decay or nodule nitrogen excretion is very small. Although the nitrogen concentration in active nodule tissue is high (often 5–6 per cent nitrogen) the proportion of nitrogen located in the nodules is rarely above about 4 per cent of the total plant nitrogen. Nitrogen in the shoots is very slowly moved to companion grasses if this transfer is effected by the normal processes of senescence, plant erosion, and decay. Passage through the grazing animal is a much more effective process. Cut and remove management systems provide little opportunity for nitrogen transfer of legume to grass, since transfer is limited to underground organs.

(d) Fertiliser nitrogen and legume/grass systems

Legume based pastures give high individual animal performance for growth, fattening, reproduction and wool growth. Cattle liveweight gain has been related positively to the proportion of legume in the sward. The consistently high protein content of legumes throughout the year is a considerable advantage over grass alone pastures, whose protein content is difficult to maintain unless very frequent nitrogen fertiliser applications are made and grazing pressure adjusted. It is reasonable to expect average animal liveweight gains of 220 kg per head per annum or 0·6 kg per head per day (550 lb or 1·3 lb per day) from large cattle breeds on good legume pastures moderately grazed. Individual stock performance on tropical legumes is usually better than or at least equal to individual stock performance on grass alone pastures fertilised with nitrogen. Some reservations might be expressed about the performance of tropical legumes for milk production, and well fertilised *Digitaria decumbens* has been superior to *M. atropurpureum* and *D. intortum* in this respect.

However, as indicated in Part 3, higher stocking rates are usually possible on grass plus nitrogen pastures than on legume based pastures. Even under very high rainfall tropical conditions, it is unusual for the year round stocking rate of a legume based pasture to exceed 4 large beasts per ha. Under the same conditions, inputs of 300 kg nitrogen per ha on a grass pasture may well support 8 beasts per ha. The high levels of production possible on a grass plus nitrogen system demand much higher inputs of fertiliser and stock than on legume based pastures, and legumes represent the better first choice for pasture development in most tropical countries.

8 Crop-pasture rotations and soil fertility

Ley farming systems in which pastures are grown in rotation with cash crops are not widely practised in many tropical countries. Farm diversification is not possible in some districts for social and economic reasons. Farm size may be too small for the added capitalisation needed and greater farm managerial skills are often needed than for a single enterprise system. However, the increased cash crop yields, the security of a diversified income, the more efficient use of crop residues possible, and the superior control of erosion make ley systems attractive to some farmers.

The chief benefit crops receive in a ley is increased nutrient availability. Table 5.14 shows the effect of various pastures grown for four years by CSIRO at Samford, Queensland (lat. 27°S, 1 150 mm rainfall). The dry matter yield of a test crop of *Sorghum vulgaris* grown after the pastures were ploughed up varied from 1 260 kg per ha on a grass (*Paspalum plicatulum*) alone treatment to 6 240 kg per ha on a grass treatment previously receiving 200 kg nitrogen per ha. This contrasted with yields of 5 520 kg per ha after *M. atropurpureum* and 1 410 kg per ha after *L. bainesii*, which added less nitrogen to the soil. Crop yield was shown to be related to soil nitrogen, and soil carbon, an index of organic matter, was also increased in the better pasture treatments. Similar beneficial effects have been reported in some African studies.

Indian studies, for instance with *Pennisetum purpureum, C. pubescens, S. guyanensis, Pueraria phaseoloides,* and *Calopogonium mucunoides* at Kanke, have also demonstrated favourable effects on following crops. Table 2.1 indicates the soil fertility changes on a humic gley soil in coastal Queensland after good pasture development.

In East Africa deep rooting grasses have brought up phosphorus and nitrogen from deep layers of the soil and have made these nutrients more available to shallower rooted crops. At Katherine, Australia (lat. 14°S, 950 mm rainfall) the reservoir of nitrate at depth under *S. humilis* pastures has been tapped by deep rooted crops of *Pennisetum americanum.*

There are some situations where pastures are not beneficial to following crops. Grass pastures may decrease nitrogen-availability, especially if there are massive amounts of carbonaceous material to be broken down. At the Brian Pastures Research Station, Gayndah, Queensland (lat. 26°S, 710 mm rainfall) yields of *Sorghum vulgaris* and of *Avena strigosa* were much higher after *Panicum maximum* or *Chloris gayana* pastures than after *Cenchrus ciliaris.* This occurred despite the greater root development of *C. ciliaris* and the decreased apparent soil density in this treatment, which is usually regarded as favourable indication. The effect was probably related to the immobilisation of nitrogen by the decomposer 'industry' working on the *C. ciliaris* roots and to the difficulty of completely removing this species. In sub-tropical districts *Medicago sativa* has had very favourable effects on subsequent cereal crops. However, *M. sativa* may decrease the growth of first year crops after ploughing if a dry season occurs at this point since *M. sativa* depletes soil moisture reserves to very considerable depths. Finally, incorrect fertilising of the ley may create subsequent crop problems. For instance, it has been found in Uganda that proper attention to the sulphur nutrition of the *S. guyanensis* ley is necessary for satisfactory yields of subsequent cash crops.

Pasture leys have a beneficial effect on soil physical structure. This has been measured in terms of improved infiltration rate, increased content of water-stable crumbs, and decreased energy inputs necessary for cultivation. These effects are usually fairly temporary in the tropics, and the main benefits are recorded in the first crop following pasture ploughing. A reduction in erosion losses

Table 5.14 Forage sorghum yield and previous pasture treatment. (*Source:* Jones, R. J. (1967). *Aust. J. Exp. Agric. Anim. Husb.* **7**:66.)

Previous Pasture Treatment	Forage Sorghum Test Crop Yield (kg/ha)	Soil Nitrogen (per cent)	Soil Carbon (per cent)
Grass Control	1 260	0·089	1·91
Grass plus 110 kg nitrogen per ha per annum	2 560	0·108	2·44
Grass plus 220 kg nitrogen per ha per annum	6 240	0·114	2·52
Grass plus *M. atropurpureum*	5 520	0·125	2·59
Grass plus *L. bainesii*	1 410	0·103	2·00

occurs under ley farming, since a well managed pasture phase is protective.

Pastures may also provide a break in the crop sequence to help control of weeds, pests and diseases.

Tropical pastures play an important part in the maintenance and building of soil fertility, and on this depends successful food production and the preservation of the human environment.

Further reading

Anderson, G. D. (1968) Effects of fertilisers on botanical composition and productivity of pasture on the sandy soils of the Tanganyika coast. *E. Afr. Agric. For. J. 34*:207.

Andrew, C. S. (1969) Problems in the use of chemical analysis for diagnosis of plant nutrient deficiencies. *J. Aust. Inst. Agric. Sci. 34*:154.

Andrew, C. S. and **Norris, D. O.** (1961) Comparative responses to calcium of five tropical and four temperate pasture legume species. *Aust. J. Agric. Res. 12*:40.

Andrew, C. S. and **Pieters, W. H. J.** (1970) Effect of potassium on the growth and chemical composition of some pasture legumes: 3 Deficiency symptoms of 10 tropical pasture legumes. *CSIRO Div. Trop. Past. Tech. Paper No. 5*

Andrew, C. S. and **Robins, M. F.** (1971) The effect of phosphorus on the growth, chemical composition, and critical phosphorus percentages of some tropical pasture grasses. *Aust. J. Agric Res. 22*:693.

Andrew, C. S. and **Pieters, W. H. J.** (1970) Manganese toxicity symptoms of one temperate and seven tropical pasture legumes. *CSIRO Div. Trop. Past. Tech. Paper No. 4*

Andrew, C. S. (1963) Copper deficiency symptoms of some tropical and temperate pasture legumes. *Aust. J. Agric. Res. 14*:654.

Barrow, N. J. (1967) Some aspects of the effects of grazing on the nutrition of pastures. *J. Aust. Inst. Agric. Sci. 33*:254.

Birch, H. F. (1960) Soil drying and soil fertility. *Trop. Agric. 37*:3.

Birch, H. F. and **Dougall, H. W.** (1967) Effect of a legume on soil nitrogen mineralisation and percentage nitrogen in grasses. *Pl. Soil. 27*:292.

Bruce, R. C. (1965) Effect of *Centrosema pubescens* Benth on soil fertility in the humid tropics. *Qld. J. Agric. Anim. Sci. 22*:221.

Bruce, R. C. (1967) Tropical legumes lift soil nitrogen. *Qld. Agric. J. 93*:562.

Bruce, R. C. (1972) The effect of topdressed superphosphate on the yield and botanical composition of a *Stylosanthes guyanensis* pasture. *Trop. Grasslds. 6*:135.

Bruce, R. C. and **Bruce, I. J.** (1972) The correlation of soil phosphorus analysis with response of tropical pastures to superphosphate on some North Queensland soils. *Aust. J. Exp. Agric. Anim. Husb. 12*:188.

Caro-Costas, R., Abruna, F. and **Vicente-Chandler, J.** (1972) Comparison of heavily fertilised pangola and star grass pastures in terms of beef production and carrying capacity in the humid mountain region of Puerto Rico. *J. Agric. Univ. Puerto Rico. 56*:104.

Chandapillai, M. M. (1972) Studies on the nodulation of *Stylosanthes guyanensis* Aubl. 1. Effect of added organic matter in four types of Malaysian soil. *Trop. Agric.* (Trinidad) *49*:205.

Davidson, R. L. (1964) Theoretical aspects of nitrogen economy in grazing experiments. *J. Br. Grassld. Soc. 19*:273.

Epstein, E. (1972) *Mineral nutrition of plants: principles and perspectives.* John Wiley.

Evans, H. J. and **Sorger, C. J.** (1966) Role of mineral elements with emphasis on the univalent cations. *Ann. Rev. Pl. Physiol. 17*:47.

Evans, T. R. (1971) Influence of nitrogen fertiliser schedule on animal production. *CSIRO Div. Trop. Past. Ann. Rept. 1970–71* p. 14.

Foster, H. L. (1971) Crop yields after different elephant grass ley treatments at Kawanda Research Station, Uganda. *E. Afr. Agric. For. J. 37*:63.

Gillard, P. (1967) Coprophagous beetle in pasture ecosystems. *J. Aust. Inst. Agric. Sci. 33*:30.

Haggar, R. J. (1971) The production and management of *Stylosanthes gracilis* at Shika, Nigeria. 1. In sown pastures. *J. Agric. Sci.* Camb. *77*:427.

Harker, K. W. (1960) Defaecating habits of a herd of Zebu cattle. *Trop. Agric. 37*:193.

Harty, R. L., Kelso, W. T., Bygott, R. B. and **Seton, D. H. C.** (1966) Effect of lucerne-wheat rotations on the yield and quality of subsequent wheat crops on the Darling Downs, Queensland. *Qld. J. Agric. Anim. Sci. 23*:261.

Henzell, E. F. (1970) Use of nitrogenous fertilisers on sub-tropical pastures in Queensland. *J. Aust. Inst. Agric. Sci. 36*:206.

Henzell, E. F. (1971) Recovery of nitrogen from four fertilisers applied to Rhodes grass in small plots. *Aust. J. Exp. Agric. Anim. Husb. 11*:420.

Henzell, E. F., Fergus, I. F. and **Martin, A. E.** (19655) Accumulation of soil nitrogen and carbon under a *Desmodium uncinatum* pasture. *Aust. J. Exp. Agric. Anim. Husb. 6*:157.

Jones, G. H. C. (1942) The effect of a leguminous cover crop in building up soil fertility. *E. Afr. Agric. J. 6*:157.

Jones, R. J. (1967) The effects of some grazed tropical grass-legume mixtures and nitrogen fertilised grass on total soil nitrogen, organic carbon, and subsequent yields of *Sorghum vulgaris. Aust. J. Exp. Agric. Anim. Husb. 7*:66.

Jones, R. J., Davies, J., Griffiths, and **Waite, R. B.** (1967) The contribution of some tropical legumes to pasture yields of dry matter and nitrogen at Samford, South-eastern Queensland. *Aust. J. Exp. Agric. Anim. Husb. 7*:57.

Jones, R. K. (1968) Initial and residual effects of superphosphate on a Townsville lucerne pasture in north eastern Queensland. *Aust. J. Exp. Agric. Anim. Husb. 8*:521.

Lee, M. T. and **Wilson, G. L.** (1972) The calcium and pH components of lime responses in tropical legumes. *Aust. J. Agric. Res. 23*:257.

t'Mannetje, L., Shaw, N. H. and **Elich, T. W.** (1963) The residual effect of molybdenum fertiliser on improved pastures on a prairie-like soil in sub-tropical Queensland. *Aust. J. Exp. Agric. Anim. Husb. 3*:20.

Masefield, G. B. (1957) The nodulation of annual leguminous crops in Malaya. *Emp. J. Exp. Agric. 25*:139.

Moore, A. W. (1962) The influence of a legume on soil fertility under a grazed tropical pasture. *Exp. J. Exp. Agric. 30*:239.

Mott, G. O. (1973) Nutrient recycling in pastures. *Proc. Amer. Soc. Agron. Symp.* 1973.

Norman, M. J. T. (1960) Performance of annual fodder crops at Katherine, N. T. CSIRO Aust. Div. Land Res. Reg. Surv. Techn. Pap. No. 9.

Norris, D. O. (1972) Leguminous plants in tropical pastures. *Trop. Grasslds. 6*:159.

Shaw, N. H. and **t'Mannetje, L.** (1970) Studies on a spear grass pasture in central coastal Queensland—the effect of fertiliser, stocking rate, and oversowing with *Stylosanthes humilis* on beef production and botanical composition. *Trop. Grasslds. 4*:43.

Sinha, A. K. and **Chatterjee, B. N.** (1966) Fertility building under grasses. *J. Br. Grassld Soc. 21*:153.

Teitzel, J. K. and **Bruce, R. C.** (1971) Fertility studies of pasture soils in the wet tropical coast of Queensland.
1. Soil-vegetation classification units. *Aust. J. Exp. Agric. Anim. Husb. 11*:71.
2. Granitic soils. *Aust. J. Exp. Agric. Anim. Husb. 11*:77.

Vallis, I. (1972) Soil nitrogen changes under continuously grazed legume-grass pastures in sub tropical coastal Queensland. *Aust. J. Exp. Agric. Anim. Husb. 12*:495.

Watson, G. A. (1959) Nitrogen fixation by *Centrosema pubescens. J. Rubb. Res. Inst. Malaya. 15*:168.

Wilson, J. R. and **Haydock, K. P.** (1971) The comparative response of tropical and temperate grasses to varying levels of nitrogen and phosphorus nutrition. *Aust. J. Agric. Res. 22*:573.

Younge, O. R. and **Plucknett, D. L.** (1975) Beef production with heavy phosphorus fertilisation in infertile wetlands of Hawaii. *Proc. IX Internat. Grassld Congr.* Sao Paulo, Brazil.

Part 6 Pasture management and productivity

1 Introduction

The usual aim of pasture management is to keep a high level of output from animals grazing pasture. This requires an understanding of both how pasture affects animal performance, and also of the effect of grazing animals on the capacity of a pasture to persist, grow and provide nutritious feed.

The efficiency of pasture production can be considered in terms of three related systems:

(a) The amount of pasture grown per unit of environmental growth factor. This may be expressed in terms of radiation energy, water, or soil nutrients.

(b) The amount of pasture eaten per unit of pasture grown.

(c) The amount of animal product obtained per unit of pasture eaten.

These three factors multiplied together determine the amount of animal output from the pasture. Each of them must be considered in developing good pasture management practice. In earlier chapters we have considered how the first factor, pasture growth, may be improved by using plant species which have a high growth potential. These make best use of the sun's energy, and of the soil moisture supply. We have also considered how the mineral nutrient deficiences of the soil may be overcome and the need for species which make best use of the nutrients available. This exercise is not worthwhile unless the additional pasture grown is eaten and converted to useful animal products.

These three factors interact with each other. Some pasture scientists have emphasised the need for lenient grazing in order to cause the pasture to grow faster, but consumption has a greater effect on production than pasture growth. A high level of consumption may reduce pasture growth rate below the maximum, so one must try to obtain a balance which results in maximum utilised pasture per unit area. For example, if a pasture grows at 80 kg/ha/day where stock consume 20 per cent of growth, pasture use is $80 \times 0{\cdot}2 = 16$ kg per ha per day. If a pasture grows at only 40 kg per ha per day where stock consume 70 per cent of growth, pasture use is $40 \times 0{\cdot}7 = 28$ kg/ha/day, a higher production figure. Consumption is closely related to the nutritive value of the pasture on offer to the grazing animal, and this partly controls the efficiency of converting pasture to animal product. The genetic capacity of animals to respond to improved forage quality and satisfactory animal husbandry also contributes to the efficiency of this conversion.

In this chapter we shall first consider how the pasture sward responds to differences in grazing and cutting practice. We need to understand how this can affect the persistence and growth rate of the pasture. We must also understand how the botanical composition, the structure, and the nutritive value of the sward can be modified by the way a producer grazes his animals or cuts the pasture to conserve feed. This leads to the question of stocking rate—the number of animals which can be safely carried—and the response of animal production to changes in forage availability. The principles behind different stocking methods—the ideas which decide the type of animals grazed and how often they are moved between different fields—are next discussed. The final sections deal with ways of overcoming seasonal feed shortages, either through pasture conservation or through the use of special purpose pastures and forage crops.

2 The response of the pastures sward to defoliation

'Defoliation' is used to mean removal of plant shoots by grazing or cutting. It is considered in terms of:

(a) Frequency—how often plant shoots are removed.

(b) Intensity—how much plant material remains after defoliation, or how much is removed.

(c) Timing—the stage of plant development and the climatic conditions at the time of defoliation.

Much research work on defoliation has been done using cutting experiments. It is not usually safe to apply these results directly to grazing practice, as cutting and grazing may have different effects. Cutting often involves more complete and sudden removal of plant tissue. Grazing involves animal treading, which is a neglected area of tropical pasture research, and the return of nutrients to the pasture as dung and urine.

(a) Effects of defoliation on the physical environment

Defoliation changes the physical environment in which the sward grows. The amount of moisture entering the soil is greater when plant cover is good. The volume of above ground parts is more important than the root mass in affecting run-off. More litter collects on the ground in leniently grazed pastures. On the other hand, although exposure of bare soil increases evaporation from the soil surface, the amount of transpiration loss depends on the amount of leaf surface present. This is most important in the early stages of pasture regrowth before a complete leaf canopy is formed over the ground surface. At Gayndah, Queensland (lat. 26°S), where frequent moisture shortages occur, pastures of *Panicum maximum* which had been cut low always grew for a longer period before wilting than pastures which had more leaf area left after cutting. In South Africa some cases have been reported where very frequent defoliation reduced the plant root system to the point where its capacity to extract water from the soil was reduced, but this rarely happens in normal grazing practice.

Heavy defoliation or pasture burning which exposes bare soil surface will cause an overall rise in soil temperatures. At the same time, there may be an increased occurrence of radiation frosts, since the soil radiates more heat at night if a pasture canopy is not present to interrupt this process, and more rapid cooling may lead to frost.

Obviously defoliation will also increase the light intensity in the lower levels of the sward. This may stimulate the germination of some light-sensitive weed seeds.

Let us now consider how pasture species differ in their tolerance of defoliation, and how pasture growth is affected by grazing or cutting.

(b) Species tolerance of defoliation

The dominance of a particular pasture species or variety in a mixed sward may be encouraged by suitable grazing or cutting practice. This results from differences in the plant material removed by defoliation and differences in the capacity of the species to form new leaves after defoliation. Another factor is the way in which one plant modifies the environment of another, e.g. by shading.

fig. 6.1 Showing crown of *Panicum maximum* eaten in preference to *Stylosanthes guyanensis* during the wet season at Borabue, Thailand

(i) Differences in intensity of defoliation

Animals grazing a mixed sward often selectively eat one species in preference to another. We cannot explain in a simple way why this occurs. Sometimes it is associated with a plant being less hairy or bearing fewer thorns than another; in other cases differences in chemical composition, e.g. in concentration of coumarin, may be the cause. The grazing history of the animals involved affect preferences, and seasonal effects can also be important. Fig. 6.1 shows selective consumption of *Panicum maximum* in preference to *Stylosanthes guyanensis* in Thailand during the wet season; in the dry season *S. guyanensis* is well accepted.

Accessibility of plant material is another factor, and differences in plant structure will affect how much is removed. The simplest example is that of species differing in height which are cut or slashed at a standard height above ground level. Obviously proportionately more of the taller species will be removed and this species will be at a disadvantage relative to the shorter species. A flowering grass has more elevated leaves and axillary buds because of the internode extension which has taken place, and is likely to suffer more removal than a non-flowering plant whose growing points are closer to ground level.

(ii) Flowering stage at which defoliation occurs

In addition to the effect on plant height, flowering may influence other factors. The timing of defoliation may allow one plant to seed but prevent another plant from seeding causing a subsequent reduction of soil seed reserves which will decrease seedling regeneration. The selective consumption by cattle of seed on the ground may also lessen future plant density. Some scientists believe that the plant is more easily damaged by a decrease in the energy level of its roots and crown if it is defoliated heavily during flowering.

Table 6.1 Cutting time and botanical change in bunch grass pasture at Gayndah, Queensland.

Time of Cutting	Species		
	Dichanthium sericeum	***Bothriochloa bladhii***	***Heteropogon contortus***
	Change in Per Cent Basal Cover		
January	**−2·9**	+1·5	+2·6
February	+0·8	**−1·9**	+0·3
March	−1·0	+2·9	**−2·9**
April	−0·4	+2·3	−1·6
May	+0·5	0·0	−0·8

These effects are illustrated in Table 6.1, which summarises the results of an experiment at Gayndah, Queensland with natural bunch grass pasture having a total basal cover of about 10 per cent (i.e. plant crowns occupied about 10 per cent of the ground). During the June to November winter and spring period these pastures were all grazed and cut in the same way. They were then closed for a hay cut taken in the January to May summer–autumn period. Table 6.1 shows the change in percentage basal cover of the three principal species which had occurred after three years treatment. It will be seen that the most unfavourable treatment for *Dichanthium sericeum* was the January hay cut, which caused a decrease of 2·9 per cent in the basal cover of this species. *Bothriochloa bladhii* was damaged most by a February cut, whilst *Heteropogon contortus* suffered the greatest deterioration after a March cut. It is interesting to note that the three species flowered in this order; *D. sericeum* usually started flowering in early January.

(iii) Removal of sites for potential regrowth

The position of the growing points is quite important in deciding how much damage will be caused by defoliation. Some scientists have classified grasses according to the distance of the main vegetative apex from the ground. They found that the plants most tolerant of heavy grazing and of fire have apices very low set, whilst plants with elevated apices only do well in lightly grazed pastures. Obviously rhizomatous plants with buried buds such as *Pennisetum clandestinum* or some forms of *Cenchrus ciliaris*, and stoloniferous plants such as *Digitaria decumbens*, have many more sites remaining for regrowth after grazing than lax plants like *Melinis minutiflora*, which tend to have elevated apices, easily removed by animals.

The legumes *Macroptilium lathyroides* and *Stylosanthes guyanensis* have few basal branches and this becomes more noticeable if there are long intervals between defoliations. *S. guyanensis* under continuous grazing at a

Table 6.2 Frequency of cutting and yield (kg/ha) *Stylosanthes guyanensis.* (*Source:* Grof. B., Harding, W. A. T. and Woolcock, R. F. (1970). *Proc. XI Internat. Grassld Congr.,* Surfers Paradise, Queensland, p. 226.)

Cutting Interval	Variety		
	cv. Endeavour	**cv. Q8442**	**cv. Schofield**
Summer:			
6 week interval	6 680	8 650	5 710
18 week interval	9 790	7 810	7 030
Winter:			
6 week interval	720	2 120	1 420
18 week interval	80	510	210

moderate stocking rate develops quite a good network of basal branching, which provides a satisfactory structure for regrowth. Table 6.2 illustrates what happens when the plant is allowed to grow unchecked for a long period and is then suddenly defoliated. This experiment was carried out at South Johnstone, Queensland (lat. 18°S) using two varieties of erect habit, Endeavour and Schofield, and a more decumbent ecotype Q 8442. In the first summer, yields of the erect varieties were slightly greater after a rest period of 18 weeks than yields obtained from a system of cuts every six weeks. However, in the following winter the 18 week cutting system showed to considerable disadvantage. This may have been associated with the basal branches growing above cutting height in the first, long 18 week rest period. Few basal buds were then left to initiate regrowth after cutting. This finding is a particularly interesting one, since for many plants the highest dry matter yields occur under a system using long intervals between cuts.

Species with a high density of growing points, e.g. *Cynodon dactylon, Pennisetum clandestinum*, are more likely to recover quickly after defoliation than species with few growing points, such as the lax *Paspalum commersonii.*

(iv) Shading of companion plants

Grazing and cutting practice reduces the opportunity tall or climbing plants have for shading shorter plants. This question has been discussed in the sections on competition in Part 2, and on legumes in Part 3. Low growing plants such as *Stylosanthes humilis, Desmodium heterophyllum*, or *Trifolium repens* are shaded by taller growing companion grasses if grazing pressure does not remove these to allow the light into the leaf canopy of the short species. As mentioned in Part 3, experimental work has been carried out with *S. humilis* and the bunch grass *Heteropogon contortus* grown together but with their roots separated by panels in the soil. This allowed the shoots to compete for light but prevented root competition for water or nutrients. Frequent defoliation which prevented the grass from shading the legume increased the percentage composition of *S. humilis* in the sward; infrequent defoliation gave *H. contortus* dominance.

Conversely, dominance of climbing or tall legumes is helped by more lenient grazing pressure. Many of these species, e.g. *Glycine wightii, Macroptilium atropurpureum* cv. Siratro, have relatively few basal buds; they escape shading by climbing up their companion grasses. Legume dominance in these circumstances is promoted by long intervals between grazing or cutting, or by low stocking rates.

Plants which have a higher growth rate under defoliation have a competitive advantage over their neighbours. They obtain an increasing share of the environmental growth factors—light, water and nutrients—which leads to increasing dominance. Superior growth rates may arise from preferential rejection by the grazing animal or from a plant structure favourable for the grazing management imposed. They may also result from growth advantages associated with the genetic constitution of the plant.

(c) Growth reactions of the defoliated sward

Despite many decades of research, the processes which control the growth reactions of the sward to defoliation are not well understood. Many different opinions can be found among scientists in this area of knowledge. Some of the misunderstandings arise because of the need to separate effects of management on pasture growth rate, and effects of management on the amount of growth utilised. For instance, the advantages of cutting or grazing close to the ground often lies in the greater proportion of material which is utilised, and may occur despite reduced pasture growth rate. Pasture material which is grown but not harvested requires an increasing amount of stem material to support it. It eventually senesces, accumulates as litter, and is broken down by soil organisms and non-domestic animals.

A usual response is that the growth of pasture is increased if the interval between grazing or cutting is long. Figure 6.2 illustrates the reaction of a *Desmodium intortum* dominant sward to different frequencies of cutting and cutting heights. The experiment was carried out at the CSIRO experiment station at Samford, Queensland (lat. 27°S). The mean annual yields of legume harvested were 4·2, 6·5 and 8·0 tonnes/ha when the sward was cut at intervals of 4, 8 and 12 weeks respectively. Percentage legume was 62, 72 and 81 in the same treatments.

Similar effects of long interval between cuts increasing yield are known to occur in grasses. Work by the CSIRO at Beerwah, Queensland (lat. 27°S) showed that the yield of *Digitaria decumbens* increased from 7·7 to 19·4 tonnes/ha as cutting interval was increased from 4 to

12 weeks. However, it should be noted that the grass dry matter produced under a lenient defoliation system is of much lower nutritive value than younger material. The need for continuity of feed presentation to the animals may also require that pastures are fed before maximum yield of dry matter is reached.

Let us now consider how defoliation affects the distribution of growth between the various plants organs, and the influence of the extent of the leaf canopy and the energy status of the sward on growth.

(i) Effect of defoliation on the distribution of growth between plant organs

The pasture material remaining after defoliation is mostly crown and stem material. As growth proceeds, the proportion of leaf increases. Frequent defoliation harvests a higher content of leaf than infrequent defoliation. This is partly because defoliation stimulates branching and leaf differentiation, and partly because the sward is maintained in a more juvenile condition. Fewer of the shoots pass through to the stage of producing much structural stem support or inflorescence material before they are removed.

Severe defoliation causes an energy shortage in the sward as the leaves which were producing photosynthate for transport to other organs of the plant are temporarily removed. The first effect of this energy shortage is seen in the growth of roots; root growth is reduced by defoliation before reduction in shoot growth is seen. In *Cenchrus ciliaris* defoliation reduces root size much more than root number. The proportion of assimilate going to the root system is decreased by defoliation. For instance, in one experiment with *Panicum maximum*, 34 per cent of growth went into the root system in a 20 day period after all leaves were removed. 52 per cent of growth went into roots if a more lenient defoliation was applied which left an LAI of 0·8.

In legumes, the root nodules are the first organs to be affected by defoliation. Heavy pasture use reduces the size which nodules attain and the number of new nodules being initiated. Drastic defoliation can cause the premature senescence of nodules and their shedding from roots. Root growth is next most sensitive; this is followed by inflorescence and stem growth, with leaf growth least affected.

(ii) Leaf area as a determinant of pasture regrowth

Some scientists argue that grazing and cutting practice should be guided by the need to achieve maximum conversion of solar radiation to plant material. This is best achieved by maintaining a leaf canopy on the sward which reduces the amount of radiation wasted by striking the soil

fig. 6.2 Yields of *Desmodium intortum* cv. Greenleaf and associate weed species (mainly *Cynodon dactylon*) as influenced by cutting at intervals of 4, 8, and 12 weeks and at 3·8, 7·5, and 15·0 cm heights. (*Source:* Jones, R. J. (1973). *Aust, J. Exp. Agric. Anim. Husb.* **13**:171.)

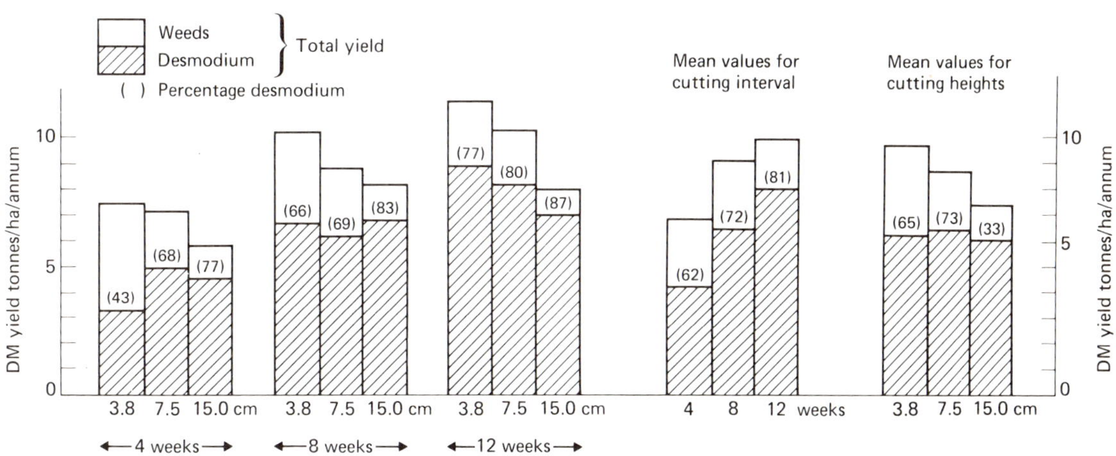

surface, and also has the best structural characteristics for maximum growth. The amount of light interception can be broadly related to the amount of leaf area present. 'Critical' LAI is defined as the amount of leaf area necessary to reduce the light value at ground level to 0·05 daylight.

Figure 6.3 shows that in a *Panicum maximum* sward, critical LAI occurred at about 3·8. Usually values of critical LAI fall in the range 3–5. It is argued that reduced growth rate occurs below critical LAI, because of inadequate light interception. Above critical LAI, growth rate will usually either stay at the same rate, or will be reduced, because of the respiratory load of the great mass of material maintained in deeply shaded conditions near the base of the sward.

It is therefore suggested that the aim of good pasture management is to maintain enough leaf surface to prevent light being wasted, and also to prevent too much build-up of plant material.

There are many factors which make it difficult to use the idea of optimum LAI in field practice. All green surfaces of the plant—leaf sheath, stem and inflorescence—contribute photosynthate. Old leaves are less photosynthetically active than young leaves, and a small canopy of mainly young leaves may give a higher sward growth rate than a larger canopy of older leaves. Light conditions within the sward are influenced by the way the leaves are arranged. Some pasture legumes such as *Macroptilium atropurpureum* cv. Siratro produce a rather narrow band of horizontal green leaves, which allows little light through to the deeper layers of the sward. A better type of leaf arrangement occurs in many grasses, in which leaf angle increases with depth, and leaves are well spaced vertically. Leaf dispersion is also important as uneven clumping together reduces growth rate.

fig. 6.3 Relationship between light value and leaf area index (LAI) of *Panicum maximum* swards. (*Source:* Humphreys, L. R. (1966). *Qld. J. Agric. Anim. Sci.* **23**:337)

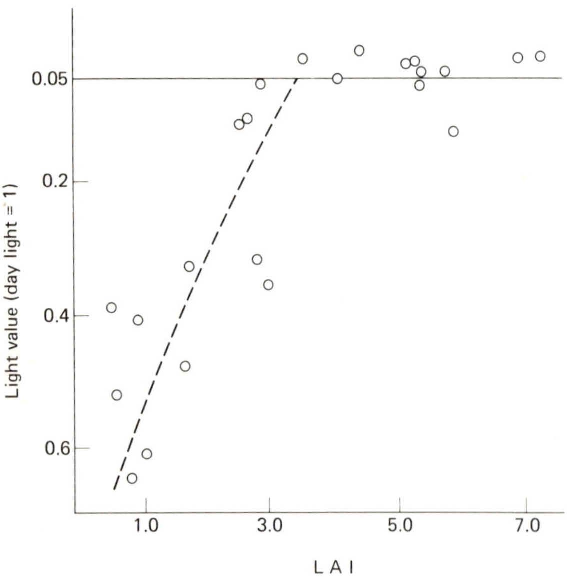

The optimum amount of leaf increases with the amount of radiation. In very sunny environments, the amount of light penetrating the lower layers of the sward is greater, and so consequently more leaf may be carried without detriment to pasture growth rate. Managing pastures to obtain efficient use of light is only possible in periods when light limits growth and in many pastoral environments these periods are of short duration. Increasing production by manipulating the leaf canopy correctly has more chance of success when the moisture supply is adequate, when fertility levels are high, and when species capable of exploiting improved environmental conditions are grown.

(iii) The energy status of the sward

The maintenance of the energy status of the roots, crown, and leaf bases of the sward is a management objective which has been given great importance in the past. It is generally accepted that if a sward is defoliated often enough and severely enough, plant death will result. A certain minimum level of energy residual is necessary to ensure survival. If the use of the pasture is so severe that insufficient labile carbohydrate is available to maintain respiration of the tissues after defoliation or during some stress period, then the plants do not survive. On the other hand, scientists do not agree about whether it is wise or not to accumulate a great deal of surplus carbohydrate for survival in the hope that this will also increase the rate of regrowth after defoliation.

Attention has been concentrated on the easily mobilisable carbohydrates, such as sugars, fructosans, dextrins, and starch. The concentration of these compounds fluctuate more widely than the structural carbohydrate components such as hemicellulose, cellulose and lignin, which have different functions. Figure 6.4

shows the pattern of accumulation of the labile carbohydrates (total available carbohydrate or TAC) in the plant organs of *Cenchrus ciliaris* swards grown at Gayndah, Queensland. These swards were cut at 5 cm (2 in.) height either at the end of winter—a treatment giving maximum opportunity for surplus carbohydrate to accumulate—or every eight weeks.

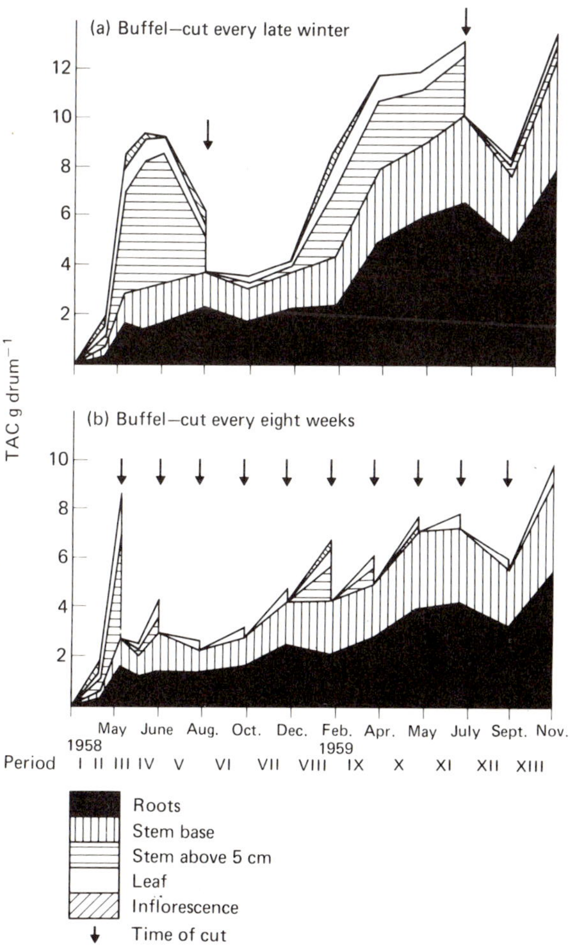

fig. 6.4 The pattern of accumulation of total available carbohydrate (TAC) in the organs of *Cenchrus ciliaris* grown under two cutting frequencies at Gayndah, Queensland. (*Source:* Humphreys, L. R. and Robinson, A. R. (1966). *Qld. J. Agric. Anim. Sci.* **23**:211)

This diagram shows a steady increase as the sward developed, in the total amount of labile carbohydrate held in the roots and stem bases. It can be seen that the steepest increases occurred in the autumn maturation period, and that little increase occurred in the summer December to February period, when pasture growth was vigorous. On the other hand, the level of labile carbohydrate in the roots and stem bases fell in the spring, with rising temperatures and the start of new growth. Frequent defoliation reduced the amount of labile carbohydrate in the roots.

After defoliation there is a fall in the labile carbohydrate level of the roots and stem bases as energy is used to maintain respiration and to start regrowth. This is followed by a rise in carbohydrate levels as growth proceeds. The decline in carbohydrate level is proportional to the severity and frequency of defoliation. Labile carbohydrate may recover balance in as little as two days after defoliation, or perhaps as long as 14 days may be needed before positive accumulation takes place.

South African work with radioactive tracer carbon has shown that leaves of *Eragrostis curvula* are exporting assimilate to the roots by the fourth day after defoliation, and that in this plant the base of the tillers constitute important energy storage organs. In *Panicum maximum* a greater movement of carbohydrate comes from the roots.

High nitrogen nutrition generally results in low levels of labile carbohydrate in the plant. More of the carbohydrate produced under high nitrogen supply is used in promoting fresh leaf growth. Under low nitrogen supply, surplus carbohydrate accumulates in the roots and the crown of the plant, since it cannot be used in promoting fresh top growth to the same extent.

Table 6.3 shows the results of an experiment which compared the relative effects of plant energy status and of leaf area on the rate of regrowth after defoliation. *Panicum maximum* swards were grown in large drums, and differences in the amount of labile carbohydrate in the roots and crown were brought about by varying management. Swards were then plucked down to different levels of leaf area remaining—zero, 0·3 LAI, and 0·8 LAI. At this time, the three carbohydrate treatments contained 0·6, 1·1, and 1·5 g per drum of labile carbohydrates in their roots and crown. Regrowth in the next 20 days was carefully measured. It will be seen from Table 6.3 that the amount of leaf surface remaining after defoliation was a greater influence on the rate of regrowth than the carbohydrate status at the time of defoliation. The rate of regrowth increased from 1·30 with zero initial LAI to

Table 6.3 Effect of initial carbohydrate status and leaf area on whole plant regrowth of *Panicum maximum*. (*Source:* Humphreys, L. R. and Robinson, A. R. (1966). *Qld. J. Agric. Anim. Sci.* **23**:211.)

Initial Leaf Area (LAI)	**Rate of Regrowth** (g/drum/day) *Initial Carbohydrate Status* (g/drum)			
	0·6	**1·1**	**1·5**	**Mean**
0·0	1·21	1·02	1·65	1·30
0·3	1·95	1·77	2·15	1·96
0·8	2·72	1·87	2·89	2·49
Mean	1·96	1·55	2·23	

2·49 g/drum/day with 0·8 initial LAI. On the other hand the highest carbohydrate level gave 2·23 g/drum/day relative to 1·96 in the lowest carbohydrate treatment, and 1·55 in the intermediate carbohydrate treatment. This sort of poor agreement between regrowth and carbohydrate status has occurred in some other investigations.

There is no doubt that the energy status of the roots and plant crown determine plant survival, and is necessary for the maintenance of perennating tissues when the plant is stressed. It is now known that compounds other than sugars, fructosans, dextrins and starch may be degraded and used in these processes; these compounds include proteins and hemicellulose. What is less clear is exactly how the energy substrate influences the rate of regrowth. Labile carbohydrates contribute to the appearance and expansion of new leaves, but these quickly become independent. The uptake of water and nutrients needs energy, and there is some evidence that good plant carbohydrate status may assist these processes, which influence the rate of leaf expansion. On the other hand, the accumulation of a massive root system of high labile carbohydrate content represents an inefficient use of energy, unless this is required for plant survival in a difficult pastoral environment.

Probably the most important objective is the maintenance of a minimum plant canopy which will not expose too much of the soil surface. Under these conditions the plant will usually be maintained in a satisfactory energy status for survival. This objective should include the maintenance of a high density of bud sites close to or below ground level; a crown structure with elevated apices is more prone to damage.

An understanding of pasture reaction to defoliation is necessary for attempts to devise management systems which will alter botanical composition in a desired way, which will adjust pasture growth rate to some specified animal needs, and which will provide the maximum use of nutrients produced by the pasture.

3 Determining the optimum stocking rate

Stocking rate is the most important management factor influencing the output of animal products from the pasture, the stability and persistence of the pasture components, and also the financial return which the farmer receives. Estimating the optimum stocking rate should therefore have the greatest importance in managerial decision-making.

Some definitions will make this discussion clearer. Stocking rate is the overall number of animals carried per unit of pasture, and the carrying capacity of a pasture is its optimum stocking rate. Stock density, on the other hand, is the number of animals per unit area in a field at any one time. Grazing pressure is the relationship between the amount of pasture present and the number of animals present. It is useful to think of the output of animal products per unit area as the product of individual animal output and the number of animals carried per unit area.

When there is a good supply of pasture, stocking rate has little effect on individual animal performance, since there is enough fodder for each animal. As the stocking rate of the pasture is increased, the individual animal output starts to decrease, as animals begin to compete with each other for the available forage supply and have less opportunity to select the most nutritious parts of the pasture. Maximum animal output per unit area occurs at a point where each animal is producing rather less than its maximum production potential. As stocking rate is increased still further, animals become so poor in condition that the additional animals carried do not compensate for the loss in individual animal production, and output per unit area decreases.

It is important to distinguish the biological production function from the economic production function. Usually the stocking rate giving maximum economic return is less than the stocking rate giving maximum biological output. This is because of the added costs in carrying many animals on a pasture. However, the first step in finding the economic optimum is to describe the biological introduction function.

The principal factors influencing the biological optimum stocking rate are:

the rate of forage growth,

the accessibility of forage to animals,

the forage nutritive value which decides the need for selective grazing to maintain animal performance,

the botanical changes and ground cover reduction induced by heavy grazing,

the seasonal variations in feed supply and opportunities for reducing these variations

the type of animal product produced.

(a) The rate of forage growth

More animals may be safely carried as the amount of pasture grown increases. Higher stocking rates can therefore be carried in areas of higher rainfall, or on soils of good fertility. If farm practices are used which increase pasture growth (e.g. sowing improved forage species, applying irrigation water, controlling woody regrowth, or using fertiliser) then there is a need to increase stock numbers if proper advantage is to be taken of the additional pasture grown.

The nature of the relationship between pasture availability and individual animal performance is illustrated

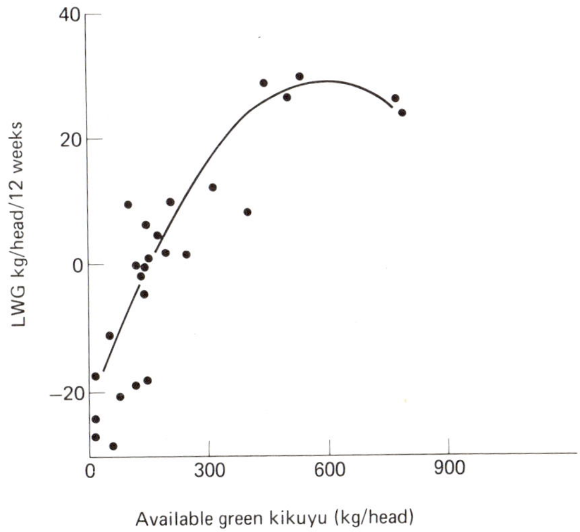

fig. 6.5 The relationship between available green kikuyu (*Pennisetum clandestinum*) (kg per head) and liveweight gain (kg per head) in autumn–winter at Wollongbar, N.S.W. (*Source:* Mears, P. T. and Humphreys, L. R. (1974). *J. Agric. Sci. Camb.* **83**:469)

Table 6.4 Production relationships between liveweight gain per animal (kg) and stocking rate (SR) as affected by nitrogen application. (*Source:* Mears, P. T. and Humphreys, L. R. (1974). *J. Agric. Sci. Camb.* **82**.)

Linear regression coefficients:	**Rate of Nitrogen Application** (kg/ha)			
	0	**134**	**336**	**672**
a	201	210	241	224
b	29·3	22·5	15·9	11·3
SR giving maximum output (beasts/ha):	3·4	4·7	7·6	9·8

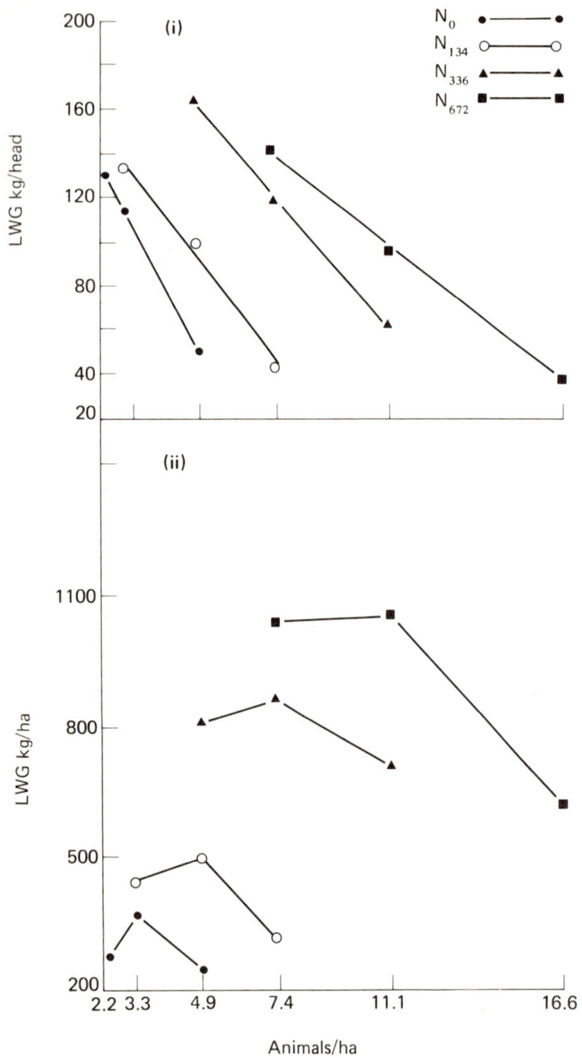

fig. 6.6 The relationship between stocking rate (animals/ha) and liveweight gain (LWG) (i) per head, (ii) per ha on *Pennisetum clandestinum* pastures at Wollongbar, N.S.W., at four levels of annual nitrogen fertiliser applications. (*Source:* Mears, P. T. and Humphreys, L. R. (1974). *J. Agric. Sci. Camb.* **83**:469)

from an experiment at Wollongbar, N.S.W. (lat. 29°S), where pastures were continuously grazed. Figure 6.5 shows that the liveweight gain per head of Angus yearlings during the autumn-winter period increased steeply as availability of green *Pennisetum clandestinum* increased. At high grazing pressures, stock lost weight. Maximum liveweight gain per head occurred when 600 kg (1 320 lb) dry weight of green *Pennisetum clandestinum* per head was available as pasture; feed availability was measured to ground level. At greater levels of feed availability no further improvement in stock growth occurred.

At Narayen, Queensland (lat. 25°S) individual animal performance depended mostly on the amount of nitrogen in the green pasture available at the beginning of each 4 weekly grazing period. Liveweight change was negative when nitrogen availability in the green pasture was less than 4 kg per head. Liveweight change reached a maximum of 0·8 kg gain per head per day when nitrogen availability was 36 kg per head. No further benefit to animal performance occurred at higher levels of nitrogen availability.

Figure 6.6 illustrates how the relationships between stocking rate and animal performance may be varied by nitrogen fertiliser use, and how the optimum stocking rate depends on the level of fertiliser input, which controls pasture growth. In this experiment four levels of ammonium nitrate were used, equivalent to annual applications of 0, 134, 336 and 672 kg N per ha per annum. Three different stocking rates were used for each nitrogen level, and these increased with increasing nitrogen level, so that similar grazing pressures were applied. The full range of stocking rates was 2·2 to 16·6 animals per ha (0·9 to 6·7 animals per ac). The data in Fig. 6.6 are taken from a rather dry year, and maximum production was only 165 kg liveweight gain per head.

The top section of the graph shows that for each nitrogen level, liveweight gain per head was linearly and negatively related to increasing stocking rate; e.g. at the N_{134} level, liveweight gain per head was 135, 101 and 43 kg for the 3·3, 5·0 and 7·5 yearlings per ha treatments respectively. At any particular stocking rate, an increase in fertiliser level caused an increase in gain per animal, e.g. at 5·0 yearlings per ha, production per head was 53, 101, and 165 kg for the N_0, N_{134} and N_{336} levels respectively. These relationships may be expressed for each nitrogen level by an equation of the simple linear regression type $y = a - bx$, where y is gain per head in kg, x is the stocking rate in animals per ha, and a and b are constants. The values of a and b are shown for each nitrogen level in Table 6.4; b represents the change in liveweight gain per head for each unit increase in stocking rate. Thus for the N_{134} level, each additional beast per ha caused an overall decrease of 22·5 kg liveweight gain per head per annum. Over two years of the experiment the average value of b for all fertiliser treatments was 18·4 kg.

Production per unit area is obtained by multiplying the stocking rate by the gain per head, and this is shown in the lower section of the graph. For each fertiliser level, the maximum liveweight gain per ha occurred at an intermediate stocking rate. Maximum recorded production per unit area was 380, 501, 873 and 1 057 kg liveweight gain per ha for the N_0, N_{134}, N_{336} and N_{672} levels respectively. If the relationships between individual animal output and stocking rate is linear, the stocking rate giving maximum animal output per unit area can be calculated using the constants from the previous equation; this stocking rate has the value $a/2b$. Thus from Table 6.4 we see that the calculated stocking rates giving maximum liveweight gain per ha were 3·4, 4·7, 7·6 and 9·8 animals per ha for the N_0, N_{134}, N_{336} and N_{672} fertiliser levels respectively.

These data show clearly how the biologically optimum stocking rate increases with additional expenditure on fertiliser; it also shows the slope of the responses obtained and the very large changes in output possible through suitable fertiliser and stocking practice. It is of course necessary to consider the long term effects on pasture stability and composition, to guarantee the conservation of an ecosystem favourable for animal production.

fig. 6.7 Showing the effect of stocking rate and superphosphate addition to a mixed pasture of *Stylosanthes humilis* and *Heteropogon contortus* at Rodd's Bay, Queensland; 0·7 beasts per ha, no superphosphate

fig. 6.8 1·1 beasts per ha, no superphosphate treatment on the same pasture as 6.7. Inferior species such as *Bothriochloa decipiens* and *Fimbristylis dichotoma* have invaded

fig. 6.9 1·2 beasts per ha, plus superphosphate treatment on the same pasture as 6.7. A good balance of grass and legume has been maintained

A second illustration is shown in Fig. 6.7, taken from a natural pasture oversown with *Stylosanthes humilis* at Rodds Bay, Queensland (lat. 24°S, 810 mm annual rainfall). In the absence of applied superphosphate, this pasture is able to support 0·7 beasts per ha. If the stocking rate is increased to 1·1 beasts per ha without fertiliser addition, the content of the perennial *Heteropogon contortus* decreases. Also less desirable species such as *Bothriochloa decipiens* and *Fimbristylis dichotoma* invade, and the proportion of bare ground increases (Fig. 6.8). However, if single superphosphate is applied at the rate of 250 kg per ha, a satisfactory cover of *S. humilis* and *H. contortus* is achieved at 1·2 beasts per ha (Fig. 6.9).

(b) Accessibility of forage to animals

Animals may not have access to all areas of pasture if animals are being grazed extensively and the property is still being developed for livestock production. The occurrence of predators or the danger of animal theft may restrict pasture use. Yarding animals at night or confining their grazing to small night paddocks causes uneven use over the farm property.

There may be insufficient watering points or sub-division fencing to ensure that all the feed grown is available for consumption. Walking distance to water varies with breed and with environmental conditions; cattle are usually considered to have an upper limit of 16 km (10 miles) to water and sheep about 5 km (3 miles) to water. In arid regions where stocking rates are low—say 0·8 sheep per ha (2 sheep per ac)—many graziers try to provide sufficient watering points so that sheep do not have to walk more than 1·5 km to water, or cattle no more than 5 km to water. Insufficient watering points lead to overgrazing, trampling damage, and pasture degradation near the water points.

(c) The nutritive value of pasture

Animals improve the quality of their diet by selectively grazing leaf and rejecting the less desirable components of the pasture. For instance, work at Serere in Uganda used surgically treated cattle, which enabled the ingested pasture to be examined after it had entered the oesophagus. This showed that the composition of *Stylosanthes guyanensis* eaten changed as grazing pressure increased. Initially animals selected leaf, then they ingested a good deal of seed and flowering material; after seven days they consumed a much higher proportion of stem.

It is quite common in the tropics to have animals grazing on grass pastures whose average nitrogen content is 0·5 per cent, but to find that animals are actually ingesting a diet of 1·3 per cent nitrogen, by selectively eating the greener tissues. *Panicum maximum* at Gayndah, Queensland, has been known to have an average nitrogen content in the shoots of 0·6 per cent during a dry frosty winter, and at the same time to bear a small amount of green leaf at the base of the plant containing 2·5 per cent nitrogen. In these circumstances, high animal performance requires that stocking rate be sufficiently low to allow the animal to graze selectively.

There is no need to provide surplus feed to permit selective grazing if forage nutritive value is high, or if animals have access to high nitrogen supplements. In such cases the optimum stocking rate is higher. Adequate fertiliser dressings will often improve nutritive value, and enable more complete pasture consumption to occur without penalty to individual animal performance. As indicated in Part 3, the nutritive value of tropical pastures is usually lower than that of the best temperate pastures, and a greater premium is placed in the tropics on the need for selectivity.

(d) Botanical composition and ground cover

Range scientists are used to thinking that non-selective grazing is necessary to keep a particular botanical composition. They will find it difficult to realise that there is need for selective grazing. Some compromise between the needs of the animal and the needs of the pasture must be made, as very heavy stocking rates usually cause the disappearance of the more valuable perennial species which are replaced by weeds. However, weeds are only important in depressing animal production if they are unpalatable or poisonous, or if the change to an annual type vegetation leads to greater fluctuations in feed availability. Too often plants are called weeds for reasons which are not connected with their known effects on animal performance.

Some pastures and soils are more easily invaded by weeds than others. *Cenchrus ciliaris* is resistant to invasion under heavily stocked conditions; *Panicum maximum* is more vulnerable in this respect. Table 6.5 shows the percentage botanical composition by weight of a mixed pasture grazed at two stocking rates. This experiment was conducted by CSIRO in the Queensland coastal lowlands at Beerwah (lat. 27°S). The more erect *Chloris gayana* was less common at the higher stocking rate (7 per cent compared with 17 per cent) where the sod-forming *Paspalum dilatatum* was the dominant grass. The scrambling legume *Desmodium intortum* only made a significant contribution at the lower stocking rate, whilst the low growing *Trifolium repens* and *Lotononis bainesii* tolerated heavy grazing.

Mention has been made previously of the shade intolerance of the annual legume *Stylosanthes humilis*, which flourishes under heavy stocking rates. Table 6.6 reports an experiment with pastures of *S. humilis* and native grass which were grazed at three rates over the summer months at Katherine, N.T., Australia (lat. 14°S, 925 mm annual rainfall). This shows a very interesting situation. Over the range of stocking rates used, liveweight gain per head was independent of stocking rate, despite the fact that the pastures were well utilised. This happened because the yield of *S. humilis* was unaffected by stocking rate.

Table 6.5 Effect of stocking rate on percentage botanical composition at Beerwah, Queensland. (*Source:* Bryan, W. W. (1970). *Proc. XI Internat. Grassld. Congr.* Surfers Paradise p. 636.)

Species	**Stocking Rate** (Animals/ha)	
	2·5	**1·2**
Chloris gayana	7	17
Digitaria decumbens	20	30
Paspalum commersonii	5	3
P. dilatatum	33	13
Desmodium uncinatum	0	1
D. intortum	1	14
Lotononis bainesii	2	1
Trifolium repens	9	5
Other species	22	16

Table 6.6 Effect of summer stocking rate on animal and pasture production at Katherine, N. T. (*Source:* Norman, M. J. T. and Phillips, L. J. (1970). *Aust. J. Exp. Agric. Anim. Husb.* **10**:710.)

	Stocking Rate (Animals/ha)		
	3·3	**2·5**	**1·7**
Liveweight gain (kg/hd)	117	118	119
Yield (kg/ha): Total pasture	2 170	2 850	3 170
S. humilis	1 640	1 920	1 840
Per cent *S. humilis*	75	67	58
Nitrogen yield (kg/ha)	41	48	49
Per cent nitrogen total pasture	1·88	1·68	1·54

fig. 6.10 *Digitaria decumbens* pastures at Mt. Cotton, south-eastern Queensland, Australia, invaded by *Cyperus* spp. and *Hypochaeris radicata* when grazed at 22 ewes per ha

fig. 6.11 The same pasture as 6.10 grazed at 15 ewes per ha

Percentage composition by weight was 75, 67 and 58 per cent in the 3·3, 2·5 and 1·7 beasts per ha treatments respectively. The average nitrogen content of the pasture on offer increased with stocking rate. It is a very happy situation when the legume balance of the pasture is improved under high stocking rates, giving a desirable botanical composition. In grazing situations with vine legumes a high stocking rate has an unfavourable effect.

Stocking rates should never be pushed to the point where soil erosion results from the absence of ground cover. Bare ground increases the amount of rainfall running off the surface of the land and causes uneven runoff. This makes stock watering more difficult and increases flooding.

(e) Seasonal variations in feed supply

In districts where annual rainfall varies greatly from year to year, the optimum stocking rate will also vary from year to year. This is not easy for a producer to predict, and many landholders will take a cautious approach to stocking rate, understocking in good years in order that sufficient feed will be available in poorer years. Variations within one year in feed supply also pose difficulties, and for very extensive stock raising enterprises the overall annual stocking rate is decided by the pasture supply available at the period of greatest forage deficiency.

The producer may employ a higher stocking rate if he has a good supply of feed sources other than those provided by normal pasture growth. Conservation of pastures or grain, purchase of feed, and the use of irrigation, can be used to reduce this problem. Their efficient use often depends upon having enough financial reserves.

Seasonal feed shortages cause more problems for graziers who run breeding stock than for graziers who conduct a mixed animal enterprise having both breeding, growing, and fattening stock. In very dry seasons it is then possible for these graziers to sell off the dry stock, but to retain a nucleus of breeding stock. Alternatively a farmer may run his enterprise mainly as a sheep property, but may buy in cattle in good years to cope with the surplus feed. He is then able to follow a more flexible stocking policy than is possible with a single enterprise system. Access to animal markets, both for ease of reducing stock numbers, and ease of restocking after a dry year, will encourage higher stocking rate.

fig. 6.12 An overgrazed stock route, reduced to bare ground

(f) Nature of animal product

The stocking rate–output relationship will depend on the sensitivity of output to nutritional stress. For sheep, wool growth is the product least reduced by stress. This is then followed by liveweight gain, reproduction, and milk production. It is therefore possible to run a higher stocking rate for wool-growing wethers than for beef producing animals. In making this assessment, meat producers are aware of the limits of liveweight gain as an index of production; the product is a saleable animal or a graded carcase. Dairy cows require a lighter stocking rate than their equivalent in beef animals on the same pasture; milk flow is very sensitive to pasture quality and greater opportunities for selective grazing are usually needed.

This discussion indicates that many factors enter the determination of the optimum stocking rate range in any farm situation. Its correct estimation will often decide the success or failure of the livestock enterprise.

4 Stocking method

(a) Type of animals grazed

Brief mention is now made of the influence of animal class and species on pasture output. Social and historical factors usually determine the types of animal used in a district, and current market requirements decide the direction of change. Flexibility of production is less easy to achieve than in the case of crop production, and with slowly reproducing animals such as buffalo or cattle, it may take some years for a policy change to be shown in output figures.

In deciding this policy, one first recognises the influence of the climatic environment on the growth and health of a specific class of stock. Second, some pasture types are more efficiently used by one class of animals than another. Thus, sheep are suited to low herbage and achieve more complete removal of plant material near the ground than cattle. There are differences between sheep breeds in their capacity to browse shrubs. Goats have very effective browsing habits and some breeds are able to climb the lower branches of trees and shrubs. Buffaloes have an excellent capacity for using low quality, fibrous roughage, whereas the quality of pasture needs to be consistently high for successful dairying. Where botanical composition is varied, animals such as sheep and cattle will graze different species preferentially. The grazing of different species of animals together in the same field can therefore often be justified by the more complete use of pasture obtained. The trampling of tall pasture by cattle will make more material accessible to sheep. Some farmers practise a system of giving their top producing animals first access to a new pasture, and following these animals with other animals of lower productive capacity, such as dry stock or sheep.

Animal species differ in their effects on pasture growth and in their patterns of pasture use. This is partly associated with differences in the way they ingest pasture; the rolling and tearing action of the ox tongue is less selective than the biting action of sheep. Horses are very selective grazers and tend to produce patches in the pasture which are heavily grazed; these patches persist from year to year. Cattle produce a finer mosaic of patches and there is less persistence of these overgrazed areas.

(b) Principles of rotational and deferred grazing

The system of grazing employed has much less influence on production than the stocking rate used. Clear demonstrations of increased stock productivity through the use of rotational grazing of tropical pastures are indeed few.

Queensland Department of Primary Industries scientists at Utchee Creek (lat. 18°S) demonstrated a small benefit from rotational grazing of *Panicum maximum–Centrosema pubescens* at high stocking rates. At their South Johnstone Research Station a few kilometres away continuous grazing was slightly superior. In one experiment at Gayndah, Queensland, the five-year mean annual liveweight gain of steers under continuous grazing was 85 kg. It was 67 kg under a rotational grazing system, and 49 kg on a more elaborate grazing system which involved resting, slashing of the pasture and reduced frequency of burning. Reports of many African studies—e.g. from Botswana, Kenya, Tanzania, and Uganda—favour continuous grazing.

It is difficult to understand why some pasture advisers, in the face of so much evidence, have continued to give a great importance to recommendations about stocking method when there are so many other factors needing attention which reliably improve production. Fencing wire can only have a small effect on pasture and animal productivity; fertiliser, improved forage species, removal of woody competitors and water supply have greater influences. It is better to assume that correct grazing pressure under continuous grazing is the normally desired system; variations of stocking method may then be introduced if these become convenient or necessary to meet specific problems.

Stocking systems are planned to meet various objectives. These may be control of botanical composition, provision for particular animal needs, improved feed use, higher pasture growth rates, or convenience of farm operations.

(i) Control of botanical composition

Deferred or rotational grazing systems have been recommended assuming that plants require a rest from defoliation. This is to complete the seeding process, to accumulate or to replenish the energy status of the plant crown and roots, to enable seedling regeneration to occur without damage by grazing stock, and to prevent the more palatable species from being eaten out. We should also remember that growth usually outstrips animal appetite at some period of the season. When pasture growth is very rapid, the presence of animals on the sward may have little influence on growth or development. Scientists do not disagree on the need to spell badly degraded pastures if restoration is to occur. However, on pasture of relatively stable botanical composition one cannot be so certain of the advantage of spelling.

Rotational grazing may be employed to meet the need to maintain a selectively grazed legume in the pasture. The maintenance of *Medicago sativa* in sub-tropical grass pastures depends on employing a rotational system. The animals are not allowed to graze for so long on one pasture that they prevent the regrowth of crowns previously eaten within the same grazing period. A system of grazing seven or fewer days on the pasture and 28 days off the pasture is preferable to one at the same stocking rate in which 14 days grazing is followed by 28 days spell.

In some pastures the persistence of a desired grass may be the problem. Table 6.7 shows the botanical composition of mixed pastures at Serere, Uganda (lat. 2°N, 1 380 mm annual rainfall, 1 110 m altitude). These were continuously grazed at 1·6, 2·5, and 4·9 beats per ha. The content of *Stylosanthes guyanensis* was independent of stocking rate, but the perennial grass *Hyparrhenia rufa* decreased from 71 to 26 per cent as stocking rate increased. At the same time, the undesirable *Sporobolus pyramidalis* increased

Table 6.7 Continuous stocking rate and botanical composition at Serere, Uganda. (*Source:* Stobbs, T. H. (1969). *Trop. Agric.* **46**:187.)

Species	**Stocking Rate** (beasts/ha)		
	1·6	**2·5**	**4·9**
Hyparrhenia rufa	71	69	26
Stylosanthes guyanensis	19	22	20
Sporobolus pyramidalis	1	2	39
Other grasses	10	8	15

from 1 to 39 per cent. A rotational system might improve the persistence of *Hyparrhenia* at high stocking rates; alternatively, a more grazing tolerant grass might be sown.

Where fire susceptible tree species require control, it is good practice to spell a field for one year in every four or five years. The rest period commences in the middle of the growing season, which allows surplus pasture growth to collect to provide good fuel for a fire at the beginning of the following wet season. The pasture is rested after the fire until the middle of the growing season, which gives the pasture species opportunity for regeneration. This system depends for its success on the occurrence of woody species which are easily destroyed by fire.

(ii) Providing for particular animal needs

Diverse grazing systems have been designed to make special provision for particular animal physiological stages and to reduce specific animal health problems. Some systems have not been tested much on tropical pastures; e.g. creep grazing systems have been devised to provide first choice of the available feed to young growing lambs or calves and to minimise intestinal parasite problems in these animals. Small gateways are constructed between fields allowing the young cattle to pass back and forth but restraining their mothers in one field.

Ectoparasites such as *Boophilus microplus* may be controlled by removing the host animals for a sufficient period to ensure the complete death of the parasite on the pasture. A three month spell in summer or a five month spell in winter may be necessary. All the cattle must be brought from the paddock to be spelled. A large accumulation of summer feed of rather low quality results from this system. Some control of internal parasites can also be obtained by spelling.

Table 6.8 Effects of continuous and rotational grazing on liveweight gain at Serere, Uganda. (*Source:* Stobbs T. H. (1969). *Trop. Agric.* **46**:293.)

	Liveweight gain (kg/ha/day) *Grazing System* Continuous	14 days on: 28 days off	7 days on: 35 days off
Dry season	1·14	1·03	0·77
Wet season	1·42	1·38	1·28

Many systems are designed to ration feed during a period of abundance so it can be used later when its effect will be most beneficial. Cows may be grazed on progressively fewer fields towards the end of their lactation period, until a block of deferred grazing material is reserved. This block might then be strip grazed as the cows approach their next calving period. Stock density is steadily decreased (by increasing the area of pasture grazed each day) to provide a rising plane of nutrition. Some attempts have also been made to adjust pasture availability to the physiological needs of the breeding ewe; special attention is given to ensuring that ewes in late pregnancy have adequate reserve feed.

Table 6.9 Rotational grazing system and cattle liveweight gain at Serere, Uganda. (*Source:* Stobbs, T. H. (1969). *Trop. Agric.* **46**:195.)

	Liveweight gain (kg/ha/day) *Grazing System* 3·5 days on: 14 days off	7 days on: 28 days off
Dry season	1·30	0·95
Wet season	2·20	2·28

Rotational or deferred grazing may be used simply to reserve feed for dry season use. At Serere, Uganda, greater animal losses occurred on continuously grazed *Panicum maximum–Macroptilium atropurpureum* pastures than under rotational grazing when a severe drought year occurred. However, the normal dry season situation (Table 6.8) showed that a six-paddock system (7 days on: 35 days off) was not so good as continuous grazing; average liveweight gains were 0·77 and 1·14 kg per ha per day respectively. Differences between grazing systems were not real during the wet seasons.

A further Serere experiment with *Hyparrhenia rufa* pastures containing either *Centrosema pubescens* or *Stylosanthes guyanensis* showed that a quick rotational grazing system (3·5 days on: 14 days off) was superior to a longer grazing interval (7 days on: 28 days off) during the dry season. Table 6·9 shows the comparative performance was 1·30 and 0·95 kg liveweight per ha per day respectively.

It is important to recognise two factors. The animal may depend upon selectivity to improve its diet if satisfactory production levels are to be maintained, and this is especially so during a dry season. Second, aged feed, which is more abundant under deferred or rotational grazing, has a lower feeding value than more recently grown material.

(iii) Improved feed use

Pastures are less easily damaged by patch grazing if a rotational system is practised. More even use of the pasture with less selectivity results. Less trampling and fouling damage occurs if strip grazing is practised. This system is best applied using an electric fence to prevent animals grazing the new feed ahead of them; a back wire behind the cattle should be used to prevent fouling or further grazing of the recently grazed areas. The animals are then kept within a narrow strip of pasture whose location is advanced each day.

fig. 6.13 *Pennisetum purpureum* used in a cut-and-carry system at Kluang, West Malaysia

The end-point in reducing least selectivity is found in a system of 'zero grazing', in which animals are not allowed on to the pasture but are kept in stalls or yards; pasture is cut and carted to the penned animals. This system eliminates fouling and trampling. Fencing, watering and shade provision are simplified and controlled feed use is possible. The system requires a lot of labour. Hand cutting of grass is common in many tropical areas and forage is collected from areas not always available to stock, e.g. strips of land near crops. Where mechanical harvesting is carried out, wet weather may interrupt the forage supply to the animals. Species resistant to cutting should be used. There are very considerable dangers of fertility transfer from the pasture to the yards and of local stream pollution; increased fertiliser use is required to maintain soil fertility.

(iv) Higher plant growth rates

Some scientists believe higher plant growth rates are achieved under rotational grazing. It is assumed that a longer interval between defoliations will provide a greater leaf canopy for an increased proportion of the year. This will in turn increase plant growth and the energy status of the sward. This question was discussed earlier in the chapter. The probable decline in nutritive value associated with long grazing intervals must also be taken into account.

Table 6.10 shows the growth of the various plants of a mixed pasture grown at Gatton, Queensland (lat. 28°S, 730 mm annual rainfall) after they had been subjected for three years to two rotational grazing systems of differing frequency. There was an increase of about 13 per cent in total yield due to lengthening the spelling period and reducing the period of grazing.

(v) Convenience of farm operations

Rotational grazing systems are convenient for some farm operations such as hay or silage making, irrigation, and herd-subdivision. It is best to separate old animals from young animals; the latter are more susceptible to internal parasites which may be transmitted from the older animals. Some graziers practise controlled seasonal mating and separate bulls from cows, except at the desired seasons. Segregation of animal classes can still be practised under

Table 6.10 Rotational grazing system and pasture yield at Gatton, Queensland. (*Source:* Yates, J. J., Russell, M. J. and Fergus, I. F. (1971). *Aust. J. Exp. Agric. Anim. Husb.* **11**:651.)

Species	**Yield** (kg/ha) *Grazing System*	
	7 days on: 21 days off	**3–7 days on: 42–63 days off**
Sorghum almum	2 250	2 760
Medicago sativa	740	870
Glycine wightii plus *Macroptilium atropurpureum*	1 080	1 240
Other species	1 000	830
Total	5 060	5 700

continuous grazing systems but some producers find it simpler to estimate and meet the changing pasture needs of particular animal groups if a more flexible system of paddocking is available.

Some indirect benefits of rotational grazing occur. Farmers who believe rotational grazing is necessary are obliged to see their animals frequently, as they are moved from field to field. This benefits stock health.

Rotational grazing should be limited to specific pastoral situations for which it is suitable. In spite of all the arguments in favour of rotational grazing, one should remember the proven farm experience that animals given maximum opportunity for selective grazing perform well, and that any system of deferment of grazing usually involves stock concentration on some other part of the farm.

fig. 6.14 Irrigated *Medicago sativa* grown for hay at Matopos, Rhodesia

5 Grazing systems involving pasture conservation

There are several reasons why pasture conservation should be practised. Pasture conservation provides more efficient animal production by making use of wasted pasture surpluses in the growing season. This reduces the decrease in body weight or production which may occur in the dry season and can prevent animal mortality in drought years. In cold climates where only limited winter grazing is available, or in Mediterranean climates where feed shortages may occur in late summer, autumn and winter, pasture conservation is accepted as normal farm practice.

In tropical regions only a very small proportion of farmers practise pasture conservation. It is used by some tropical dairy producers, by stud stock breeders, and occasionally to feed draft cattle or buffaloes, and station working or race horses. Tropical farming communities have been slow to accept pasture conservation despite quite intensive advisory programmes extending over decades in some regions. We should ask why this situation exists, and must understand the inbuilt inefficiencies in many conservation systems. We should see conservation as part of the whole farming system. Animal production is affected not only during the feeding of hay or silage but also when hay or silage is produced and after the feeding-back process has ceased.

This section discusses the various pasture and animal factors which determine the conditions under which pasture conservation is a practical, beneficial policy. A later section will discuss other approaches to reducing seasonal feed shortages.

(a) Pasture factors

(i) The extent of seasonal variation in pasture availability

Pasture conservation is only suitable where a surplus of feed occurs in one part of the year, and there is a deficiency in another part of the year. This deficiency should be expressed in terms of availability and quality of standing pasture, and not in terms of pasture growth rate during a dry season.

An experiment was carried out by the Queensland Department of Primary Industries near Atherton (lat. 17°S, 1 300 mm annual rainfall, 740 m altitude) using a high producing *Panicum maximum* var. *trichoglume–Glycine wightii* pasture. One half of the farm was rotationally grazed by dairy cows, and no conservation was practised. On the other half of the farm one half of this was conserved for silage and hay making, and cows were given access to the whole of their treatment area during the drier winter, spring and early summer months. Good quality silage was fed back when the pastures were in a dry condition. Hay was fed during the period of early summer storms when the feed available is of low bulk, and has a high water content. Both treatments were grazed at the rate of 1·25 cows per ha (0·5 cows per ac). Production was about 4 350 l milk per cow (960 gal), which was equivalent to 152 kg (335 lb) of butter fat. The animals which were fed hay and silage produced slightly less than the animals which only grazed pasture.

This is a particularly interesting result. It probably occurred because the seasonal deficiency of feed in the particular years of the experiment at this stocking rate was not sufficient to make a pasture conservation system worth while, even with a sensitive commodity such as milk.

However, if the variation in feed availability is too great, the costs of feed production and of its storage and transport may be more than the survival value of the animals. This extreme situation may occur in low rainfall areas with high rainfall variability. Selling animals at the beginning of a drought and re-stocking when good growing conditions recur may be a better policy than pasture conservation.

The scope for conservation lies somewhere between the extremes of the two cases discussed.

(ii) Effect of conservation on sward composition and vigour

A conservation cut may have a good effect on sward composition and vigour. A conservation cut will usually remove dung and urine patches, and unpalatable species will be checked. At Gayndah, Queensland, the quality of a *Panicum maximum* var. *trichoglume–Medicago sativa* sward was superior in the autumn if a conservation cut had been taken in mid-summer; the later growth had a higher nitrogen concentration.

Table 6.11 Conservation and percentage botanical composition at Townsville, Queensland. (*Source:* Ritson, J. B., Edye, L. A. and Robinson, P. J. (1971). *Aust. J. Agric. Res.* **22**:993.)

	No Conservation	Hay Cut
Stylosanthes humilis	34	43
Perennial grasses	52	33
Annual species	14	24

Table 6.11 shows the effect of three years' cutting for conservation on the composition of a pasture of *Stylosanthes humilis* and native grass near Townsville, Queensland (lat. 19°S). In the hay conservation treatment half the area was cut in the first year, and then for the next two years the whole of the area was cut, the continuously grazed cattle staying in the field. The content of the annual legume *S. humilis* was increased from 34 to 43 per cent by hay cutting, and this is considered to be beneficial. However, there was a decrease in perennial grass (mainly *Heteropogon contortus*) and an increase in annual grasses, which makes pasture production less stable. Stock production benefited a little from the conservation treatment, which was less influential than the amount of fertiliser used. Cows grew to 497 kg body weight on the non-conserved fields and to 516 kg on the conserved fields. Calving rate per cow averaged 75 and 83 per cent respectively on these treatments. In one drought year, calf birth weights were 3 kg heavier from the conserved fields, and weaning weights averaged 9 kg heavier than in the non-conserved treatments. No response to conservation occurred in the other two wetter years.

Some instances of damaging effects of conservation cuts on botanical composition occur. In semi-arid Western Queensland, the basal cover of the valuable perennial grass *Astrebla lappacea* decreases substantially after hay conservation. *Macroptilium atropurpureum* and other vine type legumes do not tolerate well a very low cut. Lengthy wet weather occurring after cutting and before the hay is collected from the field may cause the hay to rot, and some damage to pasture basal cover may result.

(iii) The quality of the pasture conserved

Many graziers do not like to store low quality roughage. For instance, at Mt. Makulu in Zambia hay was made from *Hyparrhenia* pasture. The liveweight change of cattle grazing these pastures during the June–October dry season was (a) 25 kg per head if no hay were fed, (b) 22 kg if half the area had been cut for hay, and (c) 25 kg if the whole area had been cut for hay. The hay fed had a nitrogen content of only 0·4 per cent. However, if peanut meal were fed with this low quality hay, or if urea were added to the ration, positive gains of 0·2–0·4 kg liveweight per head per day were obtained.

It is therefore necessary to decide which animal requirements must be added as a supplement, and either conserve a pasture product which will meet these standards, or else adjust the composition of the feed material with added concentrate. Work at Katherine, N.T., Australia has shown that cattle grazing grass pastures respond to supplementary feeding of high protein concentrate during the main part of the dry season. They only respond to high energy supplement at the end of the dry season when the early rains have begun.

Pasture hay is normally produced by reserving a field from grazing, cutting at a specified development stage, e.g. advanced flowering, and curing the cut material in the field until all the material has dried to a moisture content safe for storage. The material is then collected and stored. It may be mechanically picked up from the swathe and compressed into conveniently sized rectangular bales, weighing, for example, 25 kg each. These are then either stacked in the open, or placed under cover. Machines are available which produce a cylindrical bale, which rainwater runs off more easily. This is stored in the open. There is less need for covered storage if the dry season is reliably dry and the hay is to be fed in the same year. Alternatively loose (unbaled) hay may be made. Some producers in Natal take hay in under cover at a higher moisture content but they store it in stacks with large triangular vents throughout, to allow further curing in the stack. Artificial drying of pasture hay is rarely practised in the tropics.

Tropical pasture species often contain thick stems, which are much slower to dry than the leaves. By the time the stems have reached an acceptable moisture level there has been a considerable loss of quality and some leaves have fallen to the ground. This problem may be partly overcome by use of a 'conditioner'. In this the cut material is passed through rollers which crush the stem material and allow more even drying.

Pasture quality decreases steadily as the season advances, and as dry matter yield increases. In most tropical areas there is usually more fine weather for curing towards the end of the season. At this stage digestibility and nitrogen content is usually low, and fibre content is high. The uncertainty of weather conditions and the desire for better quality conserved material cut at an earlier development has led some producers to conserve pasture as silage.

The bulky, low density nature of most tropical forages makes removal of the entrapped air difficult when making silage from this material. It is therefore necessary to roll the material to make it very compact and so allow anaerobic fermentation. Chopping the pasture to short lengths im-

Table 6.12 Time of harvest, composition and silage characteristics of *Sorghum almum*. (*Source:* Catchpoole, V. R. (1972). *Trop. Grasslds.* **6**:171.)

Crop Characteristics	**Age at Harvest** (weeks)				
	4	**7**	**8**	**9**	**11**
Per cent vegetative shoots	100	43	38	18	7
Dry matter yield (kg/ha)	3 550	10 230	11 600	14 280	15 850
Per cent dry matter	9·8	19·3	21·5	25·7	31·0
Per cent nitrogen	3·0	1·9	1·8	1·5	1·4
Per cent sugars	6·4	4·9	6·5	6·6	7·2
Silage Characteristics					
Per cent dry matter	12·1	18·4	20·9	24·7	30·2
Per cent nitrogen	2·6	1·8	1·7	1·5	1·4
pH	6·4	5·7	5·2	4·6	4·7
Per cent lactic acid	0·2	0·3	1·0	4·1	2·5
Per cent volatile acids	6·0	5·2	5·3	1·9	2·5
Volatile bases (per cent total nitrogen)	29·8	33·0	21·4	10·7	11·9
Losses during storage:					
Total dry matter	19·2	8·4	4·5	6·0	5·0
Total nitrogen	29·2	12·5	10·3	7·3	8·5

proves compaction. Vacuum silage, in which the silage is made within a plastic envelope from which the air is exhausted mechanically, has become quite popular in dairying.

A second problem common to tropical forages is that sugar content tends to be low. This makes a good lactic acid fermentation harder to achieve. Work at the CSIRO Cunningham Laboratory, Brisbane has recorded 4·0–6·0 per cent water soluble sugars in *Setaria anceps*, and 3·0–3·5 per cent in *Chloris gayana*. The addition of molasses at the rate of 3·5–4·0 per cent of green weight will improve the quality of the silage made, and cutting at early heading stage is recommended. Wilting to 40 per cent dry matter content before putting in the silo gives a more stable silage.

Good quality *Glycine wightii–Panicum maximum* silage has been made at Atherton, Queensland. Nitrogen content has been above 2·0 per cent on a dry matter basis. The use of a short chopped silage and the addition of molasses at 3·5 per cent of green weight reduced the pH of silage below 4·2. At the Cunningham Ladoratory, Brisbane, *Lotononis bainesii* without molasses addition has made good silage with low content of volatile constituents. There was little loss of nitrogen or dry matter during storage, and the pH was satisfactorily low. On the other hand, *Macroptilium atropurpureum* does not make good silage without molasses addition; *Desmodium intortum* is intermediate in behaviour, and makes a satisfactory acetic acid type silage.

Sorghums and maize are normally more satisfactory materials for making silage than many other tropical forages. The effect of time of harvest on the silage characteristics of *Sorghum almum* grown near Gatton, Queensland is illustrated in Table 6.12. The crop was harvested at stages varying from 4 to 11 weeks after sowing. As development advanced, percent dry matter increased from 9·8 to 31 per cent. Dry matter yield also increased, per cent nitrogen decreased, whilst per cent sugar was relatively stable and low during this wet season. The silage produced from crops cut at 4 to 7 weeks age decomposed badly. A high proportion of nitrogen was present as volatile bases, volatile acid content was high, and lactic acid content very low. On the other hand, silage made at 9 to 11 weeks growth made reasonable quality silage with a low level of losses during storage.

(iv) Efficiency of handling and storage

It is necessary when conserving hay or silage to use physical handling and storage techniques which minimise losses during the various stages from standing pasture until the conserved material passes down the animal's throat. The nutrients present in the standing crop can be estimated and compared with the material actually ingested. Losses of up to 60 per cent have been recorded during cutting, collection from the field, transport to storage site, storage, transport to feeding site, and feeding out. However, it must be remembered that the alternatives are accepting losses as standing pasture and partially relying on animal body energy reserves for maintenance of tissues and survival.

fig. 6.15 Feeding trough at Potchefstroom, South Africa. Note absence of waste

fig. 6.16 Hereford weaners fed grass silage and urea near Marandellas, Rhodesia

(b) Animal factors

(i) Conservation for seasonal deficiencies or for animal survival during long droughts

Many farmers have found that conservation and feeding out to maintain an even feed supply throughout the year gives poor economic returns for the production of low value commodities. This is particularly so if the product is not very sensitive to nutritional stress, e.g. wool or growing stock. On the other hand, feeding of stored feed in severe droughts to maintain animal numbers may yield much better economic returns.

The nature of the animal product will influence whether seasonal feeding is profitable. There is usually a demand for fresh milk during the whole year and the rate of lactation, once depressed, is hard to restore. For fat lamb or baby beef production, where a good price may be available for rapidly grown animals, conservation will be more attractive than for wool or the production of animals in store condition.

Each situation needs to be considered in terms of local returns and costs.

(ii) Effect of pasture conservation on optimum stocking rate

Unless animals are already very underfed, conservation should enable a farmer to carry more animals safely. It is usually necessary to increase stock numbers to obtain the benefit of improved level and continuity of feed supply. At low stocking rates there will be problems in feeding substitutes. Animals fed hay or silage will simply substitute this for their intake from the grazed pasture and less of the standing pasture will be consumed. For instance, work by the Queensland Department of Primary Industries near Brisbane found that cattle supplemented with *Paspalum dilatatum* silage reduced the hours per day of grazing time. Obviously such effects are not possible if animals are fed in yards and receive their whole diet as conserved fodder.

On the other hand, a negative effect of conservation occurs if used on heavily overstocked pastures. It seems paradoxical that conservation should not assist in the situation where seasonal deficiencies most require correction. This occurs for two reasons. Firstly at high stocking rates there is less feed to be conserved. Secondly, the restriction on animal performance which occurs whilst hay or silage is being made may not be compensated for by the positive animal responses obtained when the conserved material is fed. In these circumstances conservation may actually make a seasonal feed shortage worse. Table 6.13 summarises an experiment with *Sorghum almum–Medicago sativa* pastures near Gatton, Queensland. The hay conserved was of 60 per cent digestibility. The negative effect of conservation was due to poor steer growth in the April–June period, when less feed was available, especially at the higher stocking rate.

Table 6.13 Conservation and steer production at two stocking rates near Gatton, Queensland. (*Source:* Smith, C. A. (1967). Dept. Div. Trop. Past. C.S.I.R.O. Aust. 1966/7 p. 25.)

	Stocking Rate (steers/ha)	
	2·0	**3·3**
	Liveweight gain (kg/ha/annum)	
No conservation	590	760
Hay feeding	580	670

A situation, between the two extremes discussed, occurred in experimental work at Gayndah, Queensland. *Panicum maximum* and *Medicago sativa* were used for grazing in the autumn–early summer period (May to November). The effect of mid-summer hay cut was measured on the growth of weaner cattle grazed at 2·5 and 1·5 weaners per ha. Liveweight gain per head at the higher

fig. 6.17 Afrikander cattle grazing the aftermath growth of *Chloris gayana* pastures cut for hay at Mazoe, Rhodesia

stocking rate averaged 49 kg in the control and 56 kg in the conserved treatment. At the lower stocking rate of 1·5 weaners it was 73 kg in the control and 98 kg in the conserved treatment. Thus the response to conservation was only 7 kg liveweight at the higher stocking rate, and 26 kg at the lower stocking rate, which probably approached more nearly the desirable optimum.

Work by the CSIRO near Brisbane on *Chloris gayana* or *Setaria anceps* pastures also compared output from grazing alone with output from a grazed conservation system. The average annual response to conservation was only 8 kg liveweight gain per head at 5 steers per ha, and 31 kg at 3·7 steers per ha. The response varied greatly from year to year.

Clearly there is an optimum range of stocking rate in which the best responses to conservation occur. In this range it should be possible to carry animals more safely and obtain performance at a slightly higher stocking rate than is possible without conservation. Only feed which is truly surplus to animal needs should be conserved.

(iii) Extent of compensatory weight gain

Dry season feeding levels will reduce wet season gains in proportion to the amount of gain they produced in the dry season period. Quite commonly the extent of this gain is equivalent to 0·1 to 0·5 of the additional dry season gains produced. Hence the effect of feeding back conserved pasture needs to be evaluated beyond the feeding back period into the next growing season, if a true estimate of the value of pasture conservation is to be obtained. This effect appears to arise because animals which have been given very little to eat and are then grazed on abundant pasture eat very much more than animals which have been given enough to eat throughout.

This effect is less important in young stock than in old stock, and does not apply if the animals are grown for a particular market at the end of the dry season and are sold before the compensatory weight gain period occurs. It is not relevant to breeding stock where feeding level may determine survival, the growth of the foetus, and later conception.

Pasture conservation involves a great deal of husbandry and yields less return on investment than other improvement practices such as the use of better pasture species, higher fertiliser inputs, or fodder cropping. It is a useful plan for many production situations, especially those involving high priced products grown in areas with a very dry season.

6 Year-round grazing systems involving special purpose pastures

(a) Approaches to maintaining continuity of feed supply

There are five basic approaches to the problem of keeping a supply of food available to grazing animals.

(i) Adjusting feed supply through conservation

This has been discussed in the last section.

(ii) Adjusting feed supply through purchase of feed

Purchase of concentrates is traditional for many dairy farmers. Purchase of grain and of legume hay may often be more economic than attempting to produce it on the livestock farm, especially if the environment for grain or hay production is more suitable elsewhere, or if the farm size or soils are unsuitable. Prices usually rise during periods of scarcity, and skilful purchasing of requirements in advance can save money.

(iii) Adjusting feed supply through modifying the environment

Some modifications to the environment in which plants grow can be made so that there is less variation in feed supply. The water supply available to plants may be improved by irrigation, or by cultivation and fallowing. Destruction of competing vegetation, e.g. trees and woody shrubs, will increase the amount of moisture available to the pasture. In Part 5 reference was made to the way fertiliser application can be planned so that it favours out of season growth and improves the nutritive value of standing dry season forage.

(iv) Adjusting animal requirements to the available feed supply

Grazing pressure can be easily adjusted by varying mating time. The feed requirements of the mother change throughout the year; just after weaning they may be only one-half of the level needed in early lactation. The seasonal requirements of the young progeny are similarly related to mating time. The clearest example of this is in fat lamb production, where lamb pasture requirements may be limited to a period from one month after birth to their sale about five months after birth. Some synchronisation of animal needs and pasture availability can be brought together by arranging the main calving or lambing at the beginning of the main growing season, provided other environmental factors are not unfavourable.

The purchase and sale of animals can also be used to synchronise animal needs and pasture availability. Purchase of store cattle at the beginning of the main growing season and sale of fat cattle and of barren cows before the beginning of seasonal feed shortages reduces the feed continuity problem. On the other hand, in some markets these purchase and selling seasons can cause the producer to buy animals when they are costly and sell them when they are cheap.

(v) Adjusting feed supply through addition of forage species

In Part 3 the characteristics, upon which pasture species should be judged, were discussed. Great importance was given to characteristics which assist continuity of feed supply. Species with a wide tolerance of varying climatic conditions have a longer growing season. In some environments it may be possible to grow species with complementary growth rhythms which will minimise seasonal fluctuations in output. In the Sahel and Sudanese zones of Africa the legume tree *Acacia albida* comes into leaf as the dry season begins, and sheds leaf at the beginning of the wet season. The addition of annual forage crops, may alter season of feed availability. Finally, the maintenance of good nutritive value as standing pasture in the dry season may be the most important selection characteristic.

(b) Types of feed available

Farmers usually have access to a number of feed sources depending on how many different types of farming they are carrying out and how specialised the farm is. A single farm might have available some of the following types of feed:

(i) Unimproved natural pasture

On livestock farms which are in an intermediate stage of development areas of unimproved natural pasture may occupy marshy areas, steeply sloping land, or hills with shallow skeletal soils and surface stone. These areas are all unsuited to cultivation. Natural pasture may also be cut and carted to stock from roadside verges, borders of crop areas, or other unused areas beyond the farm boundary.

(ii) Improved natural pasture

The first stage in the improvement of savanna woodland is the reduction of tree cover. Overseeding legumes such as *Stylosanthes humilis* and fertiliser application are further stages of development.

(iii) Sown legumes

Areas of legume dominant forage, e.g. *Leucaena leucocephala* may be established to supplement natural grazing.

(iv) Crop stubbles

In mixed farming areas residues from cereal and other crops may be the major source of animal diet. Many farm enterprises carry animals primarily for draft purposes, and these animals are often fed cut grass and crop material. Crop stubbles are grazed in the field in the dry season. Straw may be cut and stored, and the husks, stem and chaff collected from the threshing floor used by domestic animals. On many mixed farms in the tropics livestock feed shortage may be serious during the main growing season, when the cropping area is restricted from grazing. During the dry season the whole crop area is available for foraging.

(v) Sown pastures

Sown pastures based on grass-legume mixtures or grass plus nitrogen fertiliser are a further development stage.

(vi) Grazing crops

Rain grown grazing crops are used to meet a particular seasonal feed shortage, or because it is possible to have a high stocking rate on them.

(vii) Irrigated pastures and crops

Irrigation is not widely used in the tropics to increase livestock production, except for some high priced products such as whole milk and stud animals. Some specialised beef enterprises have developed successfully on an irrigated pasture base but these are few.

Irrigation of temperate species, e.g. *Trifolium repens* based pastures is successfully used in the sub-tropics and high altitude tropics to maintain whole-milk production in the cool season. A recent survey in southern Queensland has shown the limitations of improving the summer feed supply by growing more tropical grasses and legumes, if cows are mainly calved in winter and are at an advanced lactation stage in summer. In these circumstances milk production is very responsive to use of irrigated temperate pastures.

(viii) Pasture hay or silage

This was discussed in the last section.

(ix) Stored grain

(x) Bought concentrates

The range of possible feed combinations and proportions of feed is very wide, helping to vary the pattern of livestock and crop production according to market demands, and to manipulate the supply of feeds to meet these requirements. The three basic guiding principles are:

Farmers should grow the feeds which give the highest yields of nutrients so that the stocking rate may be increased.

Farmers should study the seasonal feed requirements of the animals they are raising, making special allowance for physiological needs such as late pregnancy and lactation.

Farmers should plan the use of a series of feeds to maintain continuity of supply, profitably employing crop residues from other enterprises.

Some additional notes are now provided on the use of forage crops, pasture legume supplements and topographical variation to assist feed continuity.

(c) Forage crops

Soil moisture availability may be reduced in time through cultivation and fallowing. On soils of good depth and water holding capacity, cultivation during the latter half of the growing season will restrict water use, by removing weeds. The moisture thus saved can then be used to grow forage crops which will be available for grazing at a time when rain grown perennial pastures are dried off.

This system has been successfully used by many landholders in southern Queensland to grow a temperate cereal, grazing oats (*Avena strigosa*), for winter and spring grazing, as discussed in Part 3. *A. strigosa* has a higher nutritive value than many tropical feeds, and good rates of fattening are achieved. It is quite usual for the crop to provide 100 days forage for weaner cattle, grazed at 4–5 weaners per ha. These young animals might gain 80 kg liveweight per head in this period, giving an output of over 300 kg liveweight per ha during this restricted period when similar performance on sown pasture is not possible.

A second reason for using forage crops is that high stocking densities are possible, especially on summer growing grass crops. However, these high stocking densities need to be considered in relation to the period of land preparation and early growth, when the crop area is reserved from stock. The excellent growth rates of *Sorghum* spp. and of *Pennisetum americanum* enable stock pressure to be taken from other pastures which may per-

Table 6.14 Time of grazing *Pennisetum americanum* and cattle performance at Katherine, N. T. (*Source:* Norman, M. J. T. and Phillips, L. J. (1968). *Aust. J. Exp. Agric. Anim. Husb.* **8**:288.)

	Date of Grazing Commencement			
	Jan 28	**Mar 24**	**Apr 29**	**May 30**
Stocking rate (beasts/ha)	3·1	4·9	4·8	5·0
Date at end of gain	June 10	June 19	July 8	Aug 13
Date at end of maintenance	June 27	Aug 3	Aug 26	Sept 2
Period of gain (days)	133	87	70	75
Period of maintenance (days)	17	45	49	20
Total period (days)	150	132	119	95
No. animal grazing (days/ha)	440	642	568	479
Gain per head (kg)	86	45	32	24
Gain per ha (kg)	236	207	152	129

form better in the cool season. Some of the highest plant photosynthetic rates in the world have been recorded from crops of *P. americanum*. At Katherine, N.T., Australia (lat. 14°S, 925 mm annual rainfall) crops grown under only moderate fertiliser regimes have averaged 13 000 kg per ha dry matter yield as described earlier.

Table 6.14 summarises a series of experiments carried out by CSIRO at Katherine on the time of grazing *P. americanum*. These crops were usually sown about December 10, and received 25 kg nitrogen per ha as ammonium sulphate and 250 kg single superphosphate per ha. It will be seen from Table 6.14 that liveweight gain decreased from 236 to 129 kg per ha as the start of grazing was delayed in stages from January 28 to May 30. A slightly lower stocking rate was used in the earliest grazing treatment, which provided forage on offer of satisfactory quality; growth was checked by consumption, and almost complete crop use was obtained. A later commencement data for grazing gave maximal or near maximal crop growth, but forage quality decreased so much with age that stock performance was inferior and crop use incomplete. The nitrogen content of the crops decreased to about 0·8 per cent by late March, and 0·5 per cent by May; the nitrogen content of heads was superior to leaves, which were better than stems. However, stock selectively consumed heads and stems, at least in the earlier phases of grazing. This may have been associated with better carbohydrate availability in these fractions.

These results do not offer much prospect of using *P. americanum* crops for cool season grazing, although better results have occurred in crops which have flowered more than the crops on which Table 6.14 is based. Leguminous forage crops, e.g. *Lablab purpureus* or *Vigna sinensis* may be expected to maintain stock production longer into the cool season, although this will be at lower stocking rates.

(d) Use of special purpose pasture legume stands

When grass dominant pastures fall in nutritive value and the nitrogen content of the pasture actually ingested falls below about 1·1 per cent (7 per cent crude protein) we can be sure that this low protein content is limiting the amount of grass eaten. Providing special purpose areas which are legume dominant and which are grazed together with the inferior quality pasture may lead to its better use. Supplementation with urea or other high nitrogen supplements has similar effects. Establishment of legumes

fig. 6.18 A drum and tank for safely feeding urea and molasses

with a longer growing season will also increase the gross forage supply at this time, and the better quality carbohydrate and minerals in the legume will improve stock performance.

Table 6.15 Supplementary *Medicago sativa* areas and cattle performance at Gayndah, Queensland.

Year	Annual Liveweight Gain (kg/head)	
	Natural Pasture	Natural Pasture with *M. sativa*
1	169	201
2	42	109
3	25	137
4	120	199
5	68	137
Mean	85	157

NB 1 kg = 2·2 lb

One good example is the complementary use of *Medicago sativa* on well-drained soils in the sub-tropics. Table 6.15 shows the results of an experiment carried out by the Queensland Department of Primary Industries at Gayndah. Two-year old steers were grazed at 0·4 beasts

fig. 6.19 A $2\frac{1}{2}$ year old Sussex × Afrikander steer gaining 0·25 kg per day during the dry season at Mazoe, Rhodesia. Natural pasture grazing is supplemented with legume silage

per ha either on natural pasture alone, or on natural pasture of which one-sixth was used for raingrown *M. sativa.* The latter was subdivided into fields and rotationally grazed; stock had continuous access to natural pasture as well. The average annual steer growth on natural pasture alone was 85 kg liveweight per head but it increased to 157 kg if complementary *M. sativa* were available. The legume benefit was greatest in the dry year 3, and least in the good year 2; the legume system helped to stabilise annual output. In every year cattle in good condition were available for slaughter at the beginning of summer, when prices are usually favourable.

The use of *M. sativa* is only suitable on certain soils as persistence problems can be great in the sub-tropics and the hazard of bloat causes many producers to lose sleep at night. Shrub legumes such as *Cajanus cajan* or *Codariocalyx gyroides* are successfully used instead. *Leucaena leucocephala* is widely grown in the tropics and sub-tropics for cattle feed, and its deep rooted habit gives a longer growing season and green leaf availability when low growing herbage is dry.

Legume dominant pastures give better stock performance as standing feed in the dry season than grass pastures. *Stylosanthes guyanensis* pastures in Laos and Thailand have prevented animal weight losses and are well accepted in the dry season. Experimental work at Shika, Nigeria in the Northern Guinea Savanna Zone (1 090 mm wet season rain) suggests that allowing 1·8 ha of natural *Hyparrhenia* grassland with 0·5 ha of mature *S. guyanensis* pasture per head gives adequate dry season grazing performance. Access to *S. guyanensis* was equivalent to feeding 0·8 kg cotton seed per head per day. Cattle on grass alone lost weight if fed less than 0·5 kg cotton seed supplement.

A great deal of research on complementary grazing systems using *S. humilis* has been carried out by CSIRO at Katherine, N.T., Australia. This is a true wet/dry climate, and the low humidity and absence of rain during the dry season stops legume deterioration. Very good stock performance is possible on a diet of *S. humilis* seed mixed with dry leaf and stem. Work on various sequences of native grassland and *S. humilis* used in both the wet and dry seasons showed that the most important factor in young cattle growth was simply the length of the period of access to *S. humilis.* The period from weaning to reaching slaughter weight varied from 20 to 43 months, according to the time spent on *S. humilis* pastures. The relationship in the first 21 months after weaning was given by the formula $y = 55{\cdot}5 + 0{\cdot}38x$, where y is the liveweight gain in kg per head, and x is the number of days (out of 630) spent on *S. humilis.* Dry season performance of animals has been similarly related to the amount of access to *S. humilis.* This can be rationed on a short-term basis.

(e) Topographical variation

Variation in topography can sometimes be used to advantage in solving seasonal production problems. For example, in the sub-tropics or high altitude tropics frost-free ridges and higher slopes may have a longer period of green feed availability than the frosted valley bottoms. A late flowering legume such as *Glycine wightii* cv. Tinaroo with a long growing season may be planted on the well drained hill slopes. Grazing is stopped during the autumn to provide increased winter use.

A second example is the stategic use of swampy areas in the dry season. *Brachiaria mutica* may be planted to provide green feed during the dry season. These areas should be fenced to prevent grazing during the main growing season so that more feed is available when animals require it most.

(f) Integration of special purpose pastures

A formula has been worked out for determining the relative areas and overall stocking rates of different special purpose pastures or crops designed to be used together. Ideally, the special purpose pasture should have a seasonal growth rhythm which perfectly complements the basic pasture. This is rarely practicable. One looks for a pasture which is superior to the basic pasture in a particular season of the year. Let us assume there is one pasture B, which is deficient for a definite period (period 1), and another pasture, A, which is superior in that deficient period 1, but which is inferior at other times of the year (period 2).

Let a_1, a_2 be the stocking rates (SR) on pasture A during periods 1 and 2 respectively.

Let b_1, b_2 be the SR on pasture B during periods 1 and 2. f is the overall SR in animals per unit area. $A + B = 1$.

To support a fixed number of animals all year round, the areas used are adjusted to the requirements of the moving group so that

$$A(a_1 - a_2) = B(b_2 - b_1)$$

A and f are determined from the formulas

$$A = \frac{b_2 - b_1}{(a_1 - a_2) + (b_2 - b_1)}$$

$$f = \frac{a_1 b_2 - a_2 b_1}{(a_1 - a_2) + (b_2 - b_1)}$$

Example:

Let us assume a grazier, in Australia, with *S. humilis*—natural grass pasture (pasture B) decided to grow some *Pennisetum americanum* (pasture A) in order to take the pressure off his legume pasture during the summer months. He estimates that a crop of *P. americanum* will carry 3·5 beasts per ha from February 1 to June 1 (period 1) but will not be grazed from June 1 to February 1 (period 2). He plans to carry 0·5 beasts per ha on his *S. humilis* pasture from February 1 to June 1, and 2·0 beasts per ha on this pasture from June 1 to February 1. What proportion of *P. americanum* should be grown, and what is the overall stocking rate?

From the data given,

$$a_1 = 3{\cdot}5, \qquad a_2 = 0$$
$$b_1 = 0{\cdot}5, \qquad b_2 = 2$$

$$A = \frac{2 - 0{\cdot}5}{(3{\cdot}5 - 0) + (2 - 0{\cdot}5)} = 0{\cdot}3 \text{ and } B = 0{\cdot}7$$

and

$$f = \frac{(3{\cdot}5 \times 2) - (0 \times 0{\cdot}5)}{(3{\cdot}5 - 0) + (2 - 0{\cdot}5)} = 1{\cdot}4$$

Thus *P. americanum* would need to be 30 per cent of the area, and the overall stocking rate would be 1·4 beasts per ha. If his farm were 100 ha, it would carry 140 cattle. During the period February 1 to June 1, 105 head would run on 30 ha *P. americanum* and the remaining 35 head would graze on 70 ha *S. humilis* pasture. The latter would then carry all 140 head over the period June 1 to February 1.

A second example is given from sub-tropical Queensland, using a sequence of native pasture, sown pasture, and forage crop. Beef cows calve in November–December and are run on native pasture at 0·3 cows per ha. Animals are weaned in May weighing 160–180 kg, and are then grazed on sown *Panicum maximum–Medicago sativa* pasture at 1·5 weaners per ha, during which time they gain about 70 kg. They are then moved to native pastures for the summer months, and from December to the following May might gain between 55 and 70 kg, grazing at 1·2 animals per ha. The final fattening sequence after May is on forage crops at the rate of 3 animals per ha. This produces an animal weighing 390–410 kg liveweight at 21–22 months.

Many choices of feeding sequences are open to farmers, and the use of available feeds together can be very difficult to organise. However, the effort and expense of growing more and better pasture will be wasted if the farmer does not skilfully arrange for animals to consume the feed and to convert it efficiently into the animal products which will sell best on his own markets.

The whole field of pasture management is very complex, and only broad guide-lines can be laid down in this book. Plant and animal behaviour vary from farm to farm, and the producer must constantly watch the reactions of his pasture and the performance of his animals to secure efficient pasture use within a stable ecosystem.

Further reading

Bryan, W. W. and **Sharpe, J. P.** (1965) The effect of urea and cutting treatments on the production of Pangola grass in south-eastern Queensland. *Aust. J. Exp. Agric. Anim. Husb.* 5:433.

Edye, L. A. *et al.* (1971) Fertility and seasonal changes in liveweight of Droughtmaster cows grazing a Townsville stylo-spear grass pasture. *Aust. J. Agric. Res.* 22:963.

Edye, L. A., Ritson, J. B. and **Haydock, K. P.** (1972). Calf production of Droughtmaster cows grazing a Townsville stylo-spear grass pasture. *Aust. J. Exp. Agric. Animal. Husb. 12*:7.

Catchpoole, V. R. (1970) Laboratory ensilage of three tropical pasture legumes—*Phaseolus atropurpureus, Desmodium intortum, and Lotononis bainesii. Aust. J. Exp. Agric. Anim. Husb. 10*:568.

Catchpoole, V. R. (1972) Laboratory ensilage of *Setaria sphacelata* cv. Nandi and *Chloris gayana* cv. Pioneer at a range of dry matter contents. *Aust. J. Exp. Agric. Anim. Husb. 12*:228.

Haggar, R. J. (1971) The production and management of *Stylosanthes gracilis* at Shika, Nigeria. II. In savanna grassland. *J. Agric. Sci. /Camb./ 77*:437.

Harvey, J. M. *et al* (1963) Influence of grazing management and copper supplementation on the growth rate of Hereford cattle in south-eastern Queensland. *Qld. J. Agric. Sci. 20*:137.

Hendricksen, R. E. (1970) The productivity under grazing of *Avena strigosa* alone and in combination with *Medicago sativa Proc. XI Internat. Grassld. Congr.*, Surfers Paradise, Australia p. 814

t'Mannetje, L. (1972) The effect of some management practices on pasture production. *Trop. Grasslds. 6*:260.

Myers, L. F. (1967) Assessment and integration of special purpose pastures. I. Theoretical. *Aust. J. Agric. Res. 18*:235.

Norman, M. J. T. (1968) The performance of beef cattle on different sequences of Townsville lucerne and native pasture at Katherine, NT *Aust. J. Exp. Agric. Anim. Husb. 8*:21.

Norman, M. J. T. (1970) Relationships between liveweight gain of grazing beef steers and availability of Townsville lucerne. *Proc. XI Internat. Grassld. Congr.*, Surfers Paradise, Australia. p. 829

Scateni, W. J. (1966) Effect of variation in stocking rate and conservation on the productivity of sub-tropical pastures. *Proc. X Internat. Grassld. Congr.*, Helsinki, p. 947.

Smith, C. A. (1961) The utilisation of *Hyparrhenia* veld for the nutrition of cattle in the dry season. II. Veld hay compared with *in situ* grazing of the nature forage, and the effects of feeding supplementary nitrogen. *J. Agric. Sci. /Camb.] 57*:311.

Steinke, T. D. (1969) The translocation of C^{14}, assimilates in *Eragrostis curvula:* an autoradiographic survey. *Proc. Grassld. Soc. S. Afr. 4*:19.

Stobbs, T. H. (1969) The effect of grazing management upon pasture productivity in Uganda. 1. Stocking rate. *Trop. Agric, Trinidad, 46*:187
2. Grazing frequency. *Ibid. 46*:195
3. Rotational and continuous grazing. *Ibid. 46*:293
4. Selective grazing. *Ibid. 46*:303.

Whiteman, P. C. (1970) Seasonal changes in growth and nodulation of perennial tropical pasture legumes in the field. II. Effects of controlled defoliation levels on nodulation of *Desmodium intortum* and *Phaseolus atropurpureus. Aust. J. Agric. Res. 21*:207.

Index of Selected Scientific Names

Note: *Stylosanthes guyanensis* is now called *Stylosanthes guianensis* and *Glycine wightii* is now called *Neonotonia wightii.*

Subject Index